T0212352

Lecture Notes in Networks and Systems

Volume 440

The series "Lecture Notes in Networks and Systems" publishes the latest developments in Networks and Systems—quickly, informally and with high quality. Original research reported in proceedings and post-proceedings represents the core of LNNS.

Volumes published in LNNS embrace all aspects and subfields of, as well as new challenges in, Networks and Systems.

The series contains proceedings and edited volumes in systems and networks, spanning the areas of Cyber-Physical Systems, Autonomous Systems, Sensor Networks, Control Systems, Energy Systems, Automotive Systems, Biological Systems, Vehicular Networking and Connected Vehicles, Aerospace Systems, Automation, Manufacturing, Smart Grids, Nonlinear Systems, Power Systems, Robotics, Social Systems, Economic Systems and other. Of particular value to both the contributors and the readership are the short publication timeframe and the world-wide distribution and exposure which enable both a wide and rapid dissemination of research output.

The series covers the theory, applications, and perspectives on the state of the art and future developments relevant to systems and networks, decision making, control, complex processes and related areas, as embedded in the fields of interdisciplinary and applied sciences, engineering, computer science, physics, economics, social, and life sciences, as well as the paradigms and methodologies behind them.

Indexed by SCOPUS, INSPEC, WTI Frankfurt eG, zbMATH, SCImago.

All books published in the series are submitted for consideration in Web of Science.

For proposals from Asia please contact Aninda Bose (aninda.bose@springer.com).

More information about this series at https://link.springer.com/bookseries/15179

Cezary Biele · Janusz Kacprzyk ·
Wiesław Kopeć · Jan W. Owsiński ·
Andrzej Romanowski · Marcin Sikorski
Editors

Digital Interaction and Machine Intelligence

Proceedings of MIDI'2021 – 9th Machine
Intelligence and Digital Interaction
Conference, December 9–10, 2021,
Warsaw, Poland

 Springer

Editors
Cezary Biele
National Research Institute
National Information Processing Institute
Warsaw, Poland

Janusz Kacprzyk
Systems Research Institute
Polish Academy of Sciences
Warsaw, Poland

Wiesław Kopeć
Polish-Japanese Academy of Information
Technology
Warsaw, Poland

Jan W. Owsiński
Systems Research Institute
Polish Academy of Science
Warsaw, Poland

Andrzej Romanowski
Institute of Applied Computer Science,
Faculty of Electrical, Electronic,
Computer and Control Engineering
Łódź University of Technology
Łódź, Poland

Marcin Sikorski
Department of Applied Informatics
in Management, Faculty of Management
and Economics
Gdańsk University of Technology
Gdańsk, Poland

This work was supported by National Information Processing Institute, al. Niepodleglosci 188b, 00-608 Warsaw, Poland, VAT EU PL5250009140.

ISSN 2367-3370   ISSN 2367-3389   (electronic)
Lecture Notes in Networks and Systems
ISBN 978-3-031-11431-1   ISBN 978-3-031-11432-8   (eBook)
https://doi.org/10.1007/978-3-031-11432-8

Preface

A significant progress in the development of artificial intelligence (AI) and its wider use in many interactive products, e.g., voice assistants, road navigation systems, and in e-commerce (chatbots or recommendation systems), are quickly transforming more and more areas of our life which results in the emergence of various new social phenomena.

Many countries have been making efforts to understand these phenomena and find answers on how to put the development of artificial intelligence on the right track to support the common good of people and societies. These attempts require interdisciplinary actions, covering not only science disciplines involved in the development of artificial intelligence and human–computer interaction, but also close cooperation between researchers and practitioners.

For this reason, the main goal of the MIDI conference is to integrate two, until recently, independent fields of research in computer science: broadly understood artificial intelligence and human–technology interaction. As from 2020, topics covered during the MIDI conference will go beyond interaction design and user experience and will also focus on more general artificial intelligence-related issues.

There is no doubt that the society is getting increasingly more aware of problems related to the implementation of solutions based on artificial intelligence. However, advances in the development of artificial intelligence technology are much faster than the process of finding answers to current ethical, social or economic dilemmas. In order for the discussion on them to result in adequate solutions, experts specializing in both the AI technology and social research are necessary. As the organizers of the MIDI conference, we are convinced that the extended formula of the event will make it an even more interesting forum to share experiences between specialists in the field of artificial intelligence and human–technology interaction.

This volume has been divided into two parts: (1) Machine Intelligence and (2) Digital Interaction. As a result, the chapters deal with such topics as weather classification, object detection, automatic translation in the scientific community. The remaining chapters of the volume refer to the matters pertaining to people's behaviors in contact with virtual reality, use of virtual reality in psychological science, and to the issues that are extremely important during the pandemic, such as remote work based on IT solutions.

We hope that this book will be an inspiration and source of valuable theoretical and practical knowledge for all readers interested in new trends and for developers of end-user IT products and services. In a world where technological solutions based on artificial intelligence are created by people for people, the final success or failure of a newly created product depends on the focus on human needs. No matter how advanced a technological solution can be, if it is not adequately linked to the lifestyle or other factors determining social behaviors of the target users, they will refuse to accept it. The history of technology development and examples of failures of technological solutions created even by the world's largest tycoons show that underestimating the importance of the social perspective of the user can result in harmful and destructive effects. It should also be emphasized that the process of acquiring knowledge about the social perspective and users' needs should start as soon as possible, already at the early stages of the process of developing techno-logical solutions, which is often stressed by the authors of the texts included in this volume.

During this year's edition of the conference, two papers were awarded Best Paper Award in memory of Professor Krzysztof Marasek. He was one of the ini-tiators of the first MIDI conference focused on the user's perspective in computer science and the co-organizer of subsequent editions. Being an outstanding scientist and engineer, an excellent specialist in the field of linguistics, voice user interfaces and voice-based interaction, he understood the importance of research on techno-logical solutions conducted from the user's perspective. We hope that this year's and subsequent editions of the MIDI conference will step up to the mark and continue the work of Professor Krzysztof Marasek.

Cezary Biele
Janusz Kacprzyk
Wiesław Kopeć
Jan W. Owsiński
Andrzej Romanowski
Marcin Sikorski

Contents

Machine Intelligence

Machine Intelligence

Weather Classification with Transfer Learning - InceptionV3, MobileNetV2 and ResNet50

Patryk Młodzianowski[(✉)] [iD]

Ignacy Mościcki's University of Applied Sciences in Ciechanów, Ciechanow, Poland
`patryk.mlodzianowski@puzim.edu.pl`

Abstract. Weather recognition is a common problem for many branches of industry. For example self-driving cars need to precisely evaluate weather in order to adjust their driving style. Modern agriculture is also based on the analysis of current meteorological conditions. One of the solutions may be a system detecting weather from image. Because any special sensors are needed, the system should be really cheap. Thanks to transfer learning it is possible to create image classification solutions using a small dataset. In this paper three weather recognition models are proposed. These models are based on InceptionV3, MobileNetV2 and ResNet50 architectures. Their efficiency is compared and described.

Keywords: Machine learning · Deep learning · Transfer learning · Image classification · Convolutional Neural Networks (CNN) · Neural network architecture · Weather classification

1 Weather Classification

The most popular methods for the automatic recognition of the current weather is the use of weather stations. They are systems equipped with specialized hardware (e.g. light intensity sensors, rain detectors, temperature sensors, humidity sensors). These systems collect very detailed data that it is reflected by their relatively high cost. A cheaper alternative may be to use the magic of machine learning which allows to build a classifier that will analyze the provided photo and be able to determine the general weather condition. This paper evaluate models built on top of three different neural network architectures.

2 Transfer Learning

Transfer learning is a machine learning technique of reusing a previously prepared model to train new one for related problem. Instead of starting the learning process from scratch, it starts with the patterns learned in solving the related task. It is widely used in image recognition because neural networks try to detect edges, shapes and then features. With transfer learning it is possible to use edge

© The Author(s) 2022
C. Biele et al. (Eds.): MIDI 2021, LNNS 440, pp. 3–11, 2022.
https://doi.org/10.1007/978-3-031-11432-8_1

and shapes detection layers of pre-trained model and then train custom feature layers. It is much faster than training model from scratch, uses less computing power and allows to train model with a relatively small amount of data. Overall concept of transfer learning shows Fig. 1.

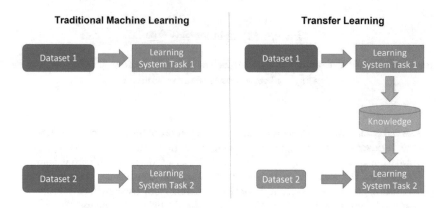

Fig. 1. Transfer learning concept

3   Architectures

Inception, ResNet, and MobileNet are the convolutional neural networks commonly used for an image classification task. Although they carry out similar problems and are based on different architectures, some differences can be expected in the results of specific tasks such as weather classification.

3.1   Inception

Inception architecture is based on two concepts - 1×1 Convolution and Inception Module. Deep neural networks are expensive in terms of computation. Thanks to 1×1 Convolution it is possible to decrease number of computations by reducing number of input channels. It causes that depth and width of neural network can be increased. Inception Module performs computations of some convolution layers simultaneously and then combines results.

InceptionV3 is a convolutional neural network that is 48 layers deep.

The network has an image input size of 299×299.

3.2   MobileNet

MobileNet targets mobile and embedded systems. This architecture is based on an inverted residual structure, which connections are between the bottleneck layers. It uses lightweight depthwise convolutions for features filtering.

This architecture allows to build lightweight models which do not need much computing power.

MobileNetV2 is a convolutional neural network that is 53 layers deep.

The he network has an image input size of 224×224.

3.3 ResNet

ResNet (Residual Networks) uses concept of identity shortcut connection that allows to jump over some layers. It partially solves vanishing gradients and mitigate accuracy saturation problem. The identity shortcuts simplifies the network and speeds learning process up.

ResNet50 is a convolutional neural network that is 50 layers deep.

The network has an image input size of 224×224.

4 Metrics

4.1 Precision, Recall, Accuracy, F1

These metrics are widely used in binary classification where only two categories are taken into consideration. In multi classification solutions they might be calculated in multiple ways, but the most popular is to calculate them as the average of every single metric across all classes. Precision represents proportion of predicted positives that are truly positive. Values closer to 1 means high precision and shows that there is a small number of false positives.

$$Precision = \frac{TruePositives}{TruePositives + FalsePositives}$$

Recall is calculated as a proportion of actual positives that have been classified correctly. Values closer to 1 means high recall and shows that there is a small number of false negatives.

$$Recall = \frac{TruePositives}{TruePositives + FalseNegatives}$$

Accuracy measures proportion of number of correct predictions to total number of samples. It helps to detect over-fitting problem (models that overfit have usually an accuracy of 1).

$$Accuracy = \frac{CorrectPredictions}{TotalPredictions}$$

F1 Score combines precision and recall metrics by calculating their harmonic mean.

$$F1 = 2 * \frac{Precision * Recall}{Precision + Recall}$$

4.2   Log-Loss, Log-Loss Reduction

Logarithmic loss quantifies the accuracy of a classifier by penalizing incorrect classifications. This value shows uncertainty of prediction using probability estimates for each class in the dataset. Log-loss increases as the predicted probability diverges from the actual label. Maximizing the accuracy of the classifier causes minimizing this function.

Logarithmic loss reduction (also called reduction in information gain - RIG) gives a measure of how much improves on a model that gives random prediction. Value closer to 1 means a better model.

4.3   Confusion Matrix, Micro-averages, Macro-averages

Confusion matrix contains precision and recall for each class in multi-class classification problem.

A macro-average computes the metric independently for each class and then take the average (treats all classes equally).

A micro-average aggregates the contributions of all classes to compute the average metric.

Micro- and macro-averages may be applied for every metric.

In a multi-class classification problem, micro-average is preferred because there might be class imbalance (significant difference between number of class' examples).

5   Dataset

Models were build on custom six class weather image dataset. Images were scraped from web. Despite the fact that the used trainer does not require data normalization [9], the images were normalized to specified aspect ratio (1:1) and size (512×512 pixels). This image size has been chosen in order to not to favor any architecture (InceptionV3 prefers 299×299, MobileNetV2 and ResNet50 prefer 224×224). Training set details are described below.

Total: 1577 images
Image format: JPEG
Image size: 512×512 pixels
Color space: sRGB
Categories:

- Clouds
- Fog
- Rain
- Shine
- Storm
- Sunrise

Table 1 presents number of images of each category and their share in total.

Table 1. Categories

Class name	Images count	Percent
Clouds	300	19%
Fog	201	12.8%
Rain	214	13.6%
Shine	252	16%
Storm	253	16%
Sunrise	357	22.6%
Total	1577	100%

6   Models

Three different models have been built to classify weather conditions. They are based on InceptionV3, MobileNetV2 and ResNet50 architectures and all of them were trained with the same dataset, specified in section above. As the solution uses transfer learning, models were trained on top of feature vectors provided by TensorFlow Hub [10–12]. All feature vectors were originally trained with ImageNet (ILSVRC-2012-CLS) dataset. Microsoft ML .NET library was used to train and evaluate models. The pipeline uses cross-validation with 10 numbers of folds. In machine learning cross-validation is a technique to measure the variability of a dataset. It also measures the reliability of any model trained using that data. Cross-validation algorithm divides randomly the dataset into subsets (folds). Then it builds a model on each subset and returns a set of accuracy metrics for each subset. One fold is used for validation and the others are used for training.

According to ML .NET trainer documentation [13], values of the hyperparameters are presented in Table 2.

Table 2. Hyperparameters

Hyperparameter	Value
Batch size	10
Epoch	200
Learning rate	0.01

What are these hyperparameters?

The batch size is a number of samples processed before the model is updated.

The number of epochs is the number of complete passes through the training dataset.

Learning rate determines the step size at each iteration while moving toward a minimum of a loss function.

The source code that have been used for model creation, training and testing is publicly available on GitHub repository [14].

The most important training metrics are shown in Table 3.

Table 3. Models evaluation metrics

Metric	InceptionV3	MobileNetV2	ResNet50
Average micro-accuracy	0.826	0.909	0.878
Average macro-accuracy	0.826	0.909	0.873
Average Log-loss	0.499	0.3	0.392
Average Log-loss reduction	0.716	0.829	0.777

Additionally each model pass performance and accuracy tests measuring total classification time for two groups of images of different size and aspect ratio. In this case there was no images normalization.

First experimental set contained 37 manually selected pictures, found on the web, of all six trained classes (clouds, fog, rain, shine, sunrise, storm). Among these data were many problematic items which poor quality have entailed that they were difficult to classify even for a human. The second set of data was "Multi-Class Images for Weather Classification" found on kaggle [15]. It had images of only four classes (clouds, rain, shine, sunrise), but was much larger than first experimental set - contained 1125 images. Both experimental datasets included images that were not used to create any of the models (Fig. 2).

(a) Prediction: fog, actual: rain   (b) Prediction: rain, actual: rain

Fig. 2. Example of potentially problematic images

Classifier trained on top of MobileNetV2 was almost 4 times faster than models based on the other architectures. This confirms that MobileNetV2 architecture would be the best choice for mobile and embedded systems. InceptionV3 and ResNet50 classified correctly more images than MobileNetV2. Table 4 and Table 5 shows time performance and accuracy for both experimental datasets.

Table 4. First experimental group - performance and accuracy

	InceptionV3	MobileNetV2	ResNet50
Elapsed time	95 s	25 s	99 s
Correct classifications	27	26	27
Incorrect classifications	10	11	10
Accuracy	0.73	0.70	0.73

Table 5. Second experimental group - performance and accuracy

	InceptionV3	MobileNetV2	ResNet50
Elapsed time	40 min 03 s	10 min 58 s	47 min 22 s
Correct classifications	1075	1098	1091
Incorrect classifications	50	27	34
Accuracy	0.95	0.97	0.96

7   Summary

Thanks to transfer learning it is possible to train custom classifiers without large
dataset and computing power. Global efficiency and accuracy may depend on
neural network architecture. All created models were able to classify weather
with accuracy of 70–73% (poor quality dataset of six classes) and 95–97% (good
quality dataset of four classes). InceptionV3 and ResNet50 architectures had
similar classification time and accuracy. MobileNetV2 had the shortest classification time and achieved competitive results. ResNet50 model achieved a slightly
higher average accuracy (based on model evaluation) than existing image-based
weather classification models [1].

Selected neural networks were compared due to significant differences in
architecture and the availability of vector features used for knowledge transfer. All vector features were trained on the same data set, which allowed to
conclude that the differences in the achieved results result mostly from architectural differences of the compared neural networks.

Weather classification is complicated due to the difficulty of extracting the
characteristics of weather phenomena. Some weather conditions are extremely
difficult for classification, e.g. heavy rain may look like a fog especially if input
image resolution is low.

Another difficulty is the possibility of mixing atmospheric phenomena. It
happens quite often that it is rainy during storm or sun shines during rain. Such
situations can not be easily resolved by simple image classification because of its
limits that is necessity to select only one option (class).

An integral part of work is a source code wrote using C# programming language and ML .NET framework [14]. Despite many searches, it was not possible
to find a program that would deal with similar issues, which would be built

with the mentioned technologies. Provided source code allows to follow line by
line whole logic being used to build models, evaluate and use them. The use of
non-mainstream technology enriches the originality of the solution.

References

1. Xia, J., Xuan, D., Tan, L., Xing, L.: ResNet15: weather recognition on traffic road with deep convolutional neural network. Adv. Meteorol. **2020**, 11 (2020). Article ID 6972826
2. Grandini, M., Bagli, E., Visani, G.: Metrics for multi-class classification: an overview (2020)
3. Zhuang, F., et al.: A Comprehensive Survey on Transfer Learning (2019)
4. An, J., Chen, Y., Shin, H.: Weather classification using convolutional neural networks, pp. 245–246 (2018). https://doi.org/10.1109/ISOCC.2018.8649921
5. Howard, A.G., et al.: MobileNets: Efficient Convolutional Neural Networks for Mobile Vision Applications (2017)
6. Zhang, Z., Ma, H., Fu, H., Zhang, C.: Scene-free multi-class weather classification on single images (2016)
7. He, K., Zhang, X., Ren, S., Sun, J.: Deep Residual Learning for Image Recognition (2015)
8. Szegedy, C., Vanhoucke, V., Ioffe, S., Shlens, J., Wojna, Z.: Rethinking the Inception Architecture for Computer Vision (2015)
9. ML.NET Trainer Characteristics. https://docs.microsoft.com/en-us/dotnet/api/microsoft.ml.vision.imageclassificationtrainer?view=ml-dotnet. Accessed 07 Nov 2021
10. Feature vector (InceptionV3). https://tfhub.dev/google/imagenet/inception_v3/feature_vector. Accessed 07 Nov 2021
11. Feature vector (MobileNetV2). https://tfhub.dev/google/imagenet/mobilenet_v2_050_224/feature_vector. Accessed 07 Nov 2021
12. Feature vector (ResNet50). https://tfhub.dev/google/imagenet/resnet_v2_50/feature_vector. Accessed 07 Nov 2021
13. ML.NET Default Hyperparameters. https://docs.microsoft.com/en-us/dotnet/api/microsoft.ml.vision.imageclassificationtrainer.options?view=ml-dotnet. Accessed 07 Nov 2021
14. Source code. https://github.com/pmlodzianowski/WeatherClassification. Accessed 07 Nov 2021
15. Second experimental dataset. https://www.kaggle.com/somesh24/multiclass-images-for-weather-classification. Accessed 07 Nov 2021

A Coarse-to-Fine Multi-class Object Detection in Drone Images Using Convolutional Neural Networks

R. Y. Aburasain[1(✉)], E. A. Edirisinghe[2], and M. Y. Zamim[3]

[1] Department of Computer Science, Jazan University, Jazan, Saudi Arabia
raburasain@jazanu.edu.sa
[2] Department of Computer Science, Loughborough University, Loughborough, UK
[3] Ministry of Education, Jazan, Saudi Arabia

Abstract. Multi-class object detection has a rapid evolution in the last few years with the rise of deep Convolutional Neural Networks (CNNs) learning based, in particular. However, the success approaches are based on high resolution ground level images and extremely large volume of data as in COCO and VOC datasets. On the other hand, the availability of the drones has been increased in the last few years and hence several new applications have been established. One of such is understanding drone footage by analysing, detecting, recognizing different objects in the covered area. In this study conducted, a collection of large images captured by a drone flying at a fixed altitude in a desert area located within the United Arab Emirates (UAE) is given and it is utilised for training and evaluating the CNN networks to be investigated. Three state-of-the-art CNN architectures, namely SSD-500 with VGGNet-16 meta-architecture, SSD-500 with ResNet meta-architecture and YOLO-V3 with Darknet-53 are optimally configured, re-trained, tested and evaluated for the detection of three different classes of objects in the captured footage, namely, palm trees, group-of-animals/cattle and animal sheds in farms. Our preliminary experiments revealed that YOLO-V3 outperformed SSD-500 with VGGNet-16 by a large margin and has a considerable improvement as compared to using SSD-500 with ResNet. Therefore, it has been selected for further investigation, aiming to propose an efficient coarse-to-fine object detection model for multi-class object detection in drone images. To this end, the impact of changing the activation function of the hidden units and the pooling type in the pooling layer has been investigated in detail. In addition, the impact of tuning the learning rate and the selection of the most effective optimization method for general hyper-parameters tuning is also investigated. The result demonstrated that the multi-class object detector developed has precision of 0.99, a recall of 0.94 and an F-score of 0.96, proving the efficiency of the multi-class object detection network developed.

Keywords: Drones · Multiclass object detection · Convolution neural networks · Unmanned aerial vehicles

1 Introduction

It is well known by the machine intelligence research community that general multi-class object detection systems have undergone a rapid development in the last few years,

C. Biele et al. (Eds.): MIDI 2021, LNNS 440, pp. 12–33, 2022.
https://doi.org/10.1007/978-3-031-11432-8_2

primarily due to the increase of use of deep machine learning based on CNNs, as has been demonstrated by the experiments conducted on the popular public databases, COCO [1] and VOC [2]. While these datasets include high-resolution images of commonly observed objects at a ground level of viewing (i.e. front, side or elevated views), the drone footage that we are utilising in the research presented in this paper have been captured looking vertically downwards from a drone flying at a fixed altitude. Hence one direct challenge this dataset poses is that none of the state-of-the art CNN networks which have been trained on the existing benchmark datasets can directly be used for detecting objects in our dataset due to the significant change of view angle, predominantly.

In literature, there have been some successful attempts in effectively using trained CNN in conducting single-class object detections in drone imagery [4, 5]. However, there has not been any attempt to develop single CNN that are capable of capturing objects of many types, i.e. multi-class object detectors. The development of multi-class CNNs that can perform multi-class object detection on drone images will require the use of more deeper CNNs and significantly more data for training such networks, as more features will have to be utilised by such networks in discriminating multiple object from each-other and from the object background. A further challenge to face in preparing data for training such networks is the need for data balancing, i.e. the need of having roughly the same number of objects for each class, in training the CNNs. In capturing data for training CNNs, this is in most cases difficult due to the relative differences of number of objects of different types physically present in captured drone footage. In the research presented in this paper, a data balancing strategy has been adopted to ensure that sufficient samples of each class are used in each iteration of training and the associated updating of the prediction loss of each iteration.

For clarity of presentation, this paper is divided into a number of sub-sections. Section 2 provides a comprehensive literature review in relation to the subject area. Section 3 shows the dataset configuration. Section 4 shows the methodology proposed for the network configuration, training, testing, evaluation, and Sect. 5 provides the experimental results and discussion. Finally, Sect. 6 concludes the research findings.

2   Related Works

Deep learning (DL) approaches outperform traditional machine learning techniques in several fields, including computer vision [6]. The history of object detection and classification using deep CNN has been motivated by its known success in general image classification tasks including but not limited to, LeNet [7], AlexNet [8], ZF-Net [9], VGG-16 [10], ResNet [11], Inception [12] and MobileNet [13].

The first attempt to apply CNN for object recognition was presented in [7]. The authors investigated the possibility of using Stochastic Gradient Descent (SGD) via backpropagation in training a CNN named the 'LeNet' for optical character recognition (OCR) in documents. This simple network comprises of two convolutional layers, a max pooling layer and a fully connected layer. Even though this architecture may not be suitable for complicated tasks, it can be considered as the backbone from which more recent, state-of-the-art CNN architectures have been derived. In the distant past, the lack of available fast computing resources resulted in the practical limitation of one's ability to train CNNs until about 2012. The period since 2012 can be considered the golden age of the application of CNNs in computer vision, as modern GPUs enhanced the capabilities to speed up the processing, especially in relation to the millions of parameters available for selection in neural networks.

The authors in [8] presented AlexNet, a CNN designed for improving classification in the ImageNet dataset [14]. This CNN comprises of five convolutional layers followed by max pooling. A dropout technique is used to reduce the overfitting. The authors investigated how increasing the number of convolutional layers improves the feature extraction in comparison with the LeNet architecture. A modified version of AlexNet was presented by [9] and named the ZF-Net. Compared to AlexNet, in ZF-Net, the filter size was reduced from 11 to 7 and the stride was reduced from 4 to 2. This has a significant impact on extracting more reliable features in the early layers. In addition, a few convolutional layers were added to improve the feature extraction.

With the idea of further deepening the CNN architecture, VGG-16 net, was proposed by [10], having a total of 16 layers. The network comprises of 12 convolutional layers and a 3 × 3 filter is used in all layers. The VGG-16 net resulted in an excellent performance as compared to the performance of previous networks. The detailed research presented in [10] has had a significant impact on the CNN community. It confirmed that increasing the depth of the model has a crucial impact on improving performance. The authors also compared the performance between VGG-16 and VGG-19, where two convolutional layers were added to the latter network. A slight improvement in the top 5% errors was obtained with VGG-19 at 8.0, as compared to 8.7 obtained with VGG-16. Nevertheless, VGG-16 is still more popular with the CNN research community as practitioners have noted its comparable accuracy with the latter, VGG-19.

The Inception-V1 architecture also called as the GoogleLeNet was presented in [15]. It aims to solve the computation costs of the very deep CNN by applying a 1 × 1 convolution and concatenating the channels. There are 27 layers, of which 22 are convolutional. Rather than using the fully connected layer before the softmax layer, this architecture uses global average pooling towards the end of the network. Using these two techniques successfully reduces the number of parameters from 12M in VGG-16 to 2.4M. This significantly speeds up the process and an illustration of the difference between the fully connected layers and the global average pooling can be found in [5].

However, increasing the depth of CNNs used, by only stacking the layers, leads to the gradient to vanish or explode, consequently increasing the training errors, more details can be found in [11]. To solve the problem of the need for a deep net without a vanishing gradient, in [11] the Microsoft community presented ResNet. This architecture has 152 convolutional layers, which is substantially higher than before. The idea here is that by applying the so-called *shortcut-connection* and building up residual blocks means that the ensemble of the smaller networks can benefit from the very deep net, without overfitting the model. Since its introduction, this architecture has become very popular within the CNN research community. It significantly improves the object detection and segmentation accuracy achievable using CNN.

Further versions of the original Inception network (i.e. V1) have been released subsequently, namely V2, V3, and V4. The former two were introduced in [16], while the latter was introduced in [12]. In Inception V2, a smart factorisation method is used, aiming to improve the computational complexity. In Inception V3, an RMSProp optimiser, batch normalisation, and adding label smoothing to the loss function are added without drastically changing the architecture. Furthermore, Inception V4 [12] combines Inception V1 with ResNet, which significantly benefits from a very deep network architecture but yet minimises the number of parameters.

A trading off between the speed and accuracy has been proposed in the design of a CNN architecture specifically proposed for applications within the mobile and embedded systems, named MobileNet in [13]. The main idea here is the adaptation of depth-wise separable convolutions for all convolutional layers, except the first layer, which is fully convolutional. This architecture demonstrated a competitive accuracy and benefits when working specifically with limited resource.

Moreover, deep CNNs has been proven to outperform conventional machine learning methods in object detection tasks [6]. Several applications have attracted the attention of both practitioners and academics, who have effectively deployed CNNs for video surveillance, autonomous driving, rescue and relief operations, robots in industry, face and pedestrian detection, understanding UAV images, recognising brands and text digitalisation, among others.

Object detection using CNNs is basically an extension of the meta-architectures used in general feature extraction and classification tasks. For example, LeNet [7], VGG-16 [10], ResNet [11], and Inception [12] are popular feature extraction architectures. The object detection is performed by expanding these architectures with more layers responsible for object detection. There are two approaches in the used in object detection domain; two-stage object detection as in [17–19] and Single-shot object detector as in [20] and [21].

In 2014, *R-CNN* object detection architecture was shown to improve the detection of objects in the VOC2012 dataset [2] in comparison to conventional machine learning techniques through combing Selective Search [22] with CNNs used for classification [17]. The authors proposed the use of Selective Search to generate 2000 region proposals and use 2000 CNNs for classification. However, this design led to the increase of algorithmic complexity and time consumption, even though the object detection accuracy and has been improved. In *Fast R-CNN* [18], a single CNN is used which significantly reduces the time consumption. The last version of this series called *Faster R-CNN* which replaces the Selective Search with a Region Proposal Network (RPN) for proposing regions. on the other hand, sharing the computation over every single convolutional layer was conducted for R-FCN when [23] proposed recognising the object parts and location variance using a position-sensitive score maps strategy. This method delivers higher speeds when compared with Faster R-CNN, with comparable accuracy.

The second approach in object detection uses a single CNN for detecting objects using the raw pixels only (i.e. not pixel regions) as in the case of the Single-Shot Detector (SSD) proposed in [20] and You Only Looks Once (YOLO) proposed in [21]. The latter approach outperforms the former in detecting objects in many benchmark datasets, including COCO [1] and VOC2012 [2] datasets, beside its capability to reduce complexity and time consumption. While SSD uses VGG-16 [10] for feature extraction, YOLO uses a custom architecture called *Darknet-19* [24]. Neither of these approaches used the filtering steps that ensure each location has a minimum probability of having an object.

YOLO has been proven to be one of the most efficient object detection architectures specifically suitable for real time applications [21]. It uses a custom meta architecture based on Darknet [24] for feature extraction. The first version of YOLO, i.e., V1, is simple in nature when compared to the subsequent two versions. It consists of 24 convolutional layers followed by fully connected layers [21]. In YOLO-V2, batch normalization and anchor boxes for bounding box prediction is utilised with the aim to improve the localisation accuracy. Significant improvement in accuracy is achieved in YOLO-V3. This architecture is changed by increasing the convolutional layers to 106, building residual blocks and skipping the connection to improve the detection at different scales. Also, they change the square errors in the loss function to cross-entropy terms and replace the softmax layer with a logistic regression that predicts the label by giving a threshold value. However, [20] showed that SSD performs better than YOLO-V1 and V2 because the predicted boxes for each location are higher than in the first two versions of YOLO. However, YOLO-V3 [25] outperforms SSD in several datasets, including the benchmark COCO dataset.

On the other hand, there are a considerable number of attempts in applying CNN object detection architectures on drone-based imagery system. Generally, the approaches proposed in literature can be categorised into three different application areas based on their purpose of use, namely, obstacle detection for ensuring safe-flying of the drones, creating DNN models for embedding within a drone's hardware and object detection/recognition/localization for aerial monitoring and surveillance of large areas [26]. The latter category with wide application areas and considerable open research problems is the focus area of the research conducted in this research. Few attempts has been published to detect different objects in drone footage using CNNs based learning as in [27–30] and [31], etc. The binary classifiers for the detection of palm trees, as one of the objects in the designed multiclass detector, can be found in [32–37] and [37, 38] and [39]. For more details we refer the readers to [5]. Furthermore, there are a few researches discussed the animal detection as in [31, 40–43] and [4].

Most previous work applied CNNs to detect objects of multiple classes as in [44] and [45] in applications related to the detection of obstacles in the flight path of a drone, thereby addressing the drone's safe flying. In contrast analysing drone footage to detect and classify multiple objects is rarely studied. To the best of our knowledge, the few published attempts recovered via the literature review conducted are detailed below.

For infrastructure assessment and monitoring purposes of the electric power distribution industry, the authors in [29] proposed the detecting three different classes of objects using a single CNN, namely power lines/cables, pylons and insulators, from drone images, with the aim of automatic maintenance and insurance purposes. The authors investigated the use of the pre-trained CNN model GoogleLeNet and fine-tuned it with their dataset before they applied Spectral Clustering [46] for further improvement of the results obtained.

Moreover, automatic railway corridor monitoring and assessment by using DNNs on images captured by drones was proposed by [27]. The pre-trained GoogleLeNet architecture and an architecture proposed by the authors were re-trained and trained respectively to detect and classify five different classes of objects, namely, lines, ballast, anchors, sleepers and fasteners. It was shown that when using the novel architecture proposed the F-score reduced from 89% to 81% as compared to using GoogleLeNet at a ten-fold reduction of network parameters due to the simpler architecture of the proposed network. With a similar focus in mind, the authors of [47] minimised the number of convolutional layers of their proposed architecture and obtained significantly good results in multi-class object classification in the area of using robots in detecting threats in crises and emergency situations. Further, the authors in [48] discussed how integration of CNN technology can be addressed in small drones by applying transfer learning and saving only the last few layers of the CNN to enable embedding the system into a drone's cameras for autonomous flight.

To conclude, there is no previous attempts of investigating the applicability of CNNs in detecting objects in drone-based imagery system with the specification given in this research dataset as shown in Sect. 3. This research reveals the significant of the number of convolutional layers, pooling type in the pooling layer, learning rate and the optimization method in improving the multi-class detector in drone-based images which is the main contribution of this research.

3 Dataset Configuration

The research dataset comprises of 221, large aerial view images of size 5472×3648 pixels, captured by a drone. An example of such an image is illustrated in Fig. 1. However, for the purpose of conducting the research proposed in this paper, three objects have been labelled: 'palm trees', sheds and 'group-of-animals'. The labelled data that consists of the above three classes, will then be used for training the multi-class object detector. The combined dataset consists of 900 images of size 416×416. This dataset has 1753, 3300 and 3420 bounding boxes for the palm tree, group of animals and sheds respectively.

3.1 Data Balancing Strategy

An important preparation step in training a multi-class model using min-batch gradient descent approach to minimise its loss function is to ensure a balanced number of labelled objects for each class, per iteration. This is because, to ensure slopping the loss function toward the minimum (weight updating), the presence of an inadequate number of samples for a particular class complicates the training process; increases noise and practically results in a high bias. This reflects the need for increasing the number of training samples.

The number of bounding boxes in the multi-class dataset has group-of-animals (3300) and sheds (3420) objects and a significantly low number of palm-tree objects (1753). The majority of raw images captured by the drones used for data collection in our experiments were of animal or crop farming areas. The presence and spread of palm trees in such farms were sparse and collecting sufficient samples of palm free was therefore difficult. Further due to the significant within-class variations of group-of-animals (or sheds), if

Fig. 1. A sample of the drone-based desert image dataset [3]; image dimensions, 5472 × 3648 pixels.

one is to attempt developing a CNN network for detection group-of-animals, a large number of such objects will be required in training, testing and validation. Given the above, 150 further images of size 500 × 500 (different magnitude) were cropped from the raw, large sized images captured by the drones. The idea is to use these additional images to boost the number of samples needed in a particular class, in the process of balancing data. Following this the total number of palm trees available for training has been increased to 2271 from the original 1753. The combined dataset used in training YOLO-V3 for multi-class object detection therefore have has 1050 cropped images in total, divided as 85% for training and 15% for testing. Practically, all the images are saved in a single folder and named with *n* number of names, which is the number of classes. This is to ensure the division of the training and testing set using the determined percentage of each class. With this strategy, the palm tree training samples are balanced, and this ensures a sufficient number of palm tree bounding boxes are trained per iteration. The final research dataset used in this research is shown in Table 1.

Table 1. The multi-class object detection dataset after data balancing strategy (# refers to 'Number of').

	#Images	#Palm tree boxes	# Groups-of-animal boxes	#Sheds boxes
Palm tree Dataset	350	**1675**	260	320
Group-of-Animals Dataset	350	320	**1760**	1410
Sheds Dataset	350	276	1280	**1600**
Multi-class Dataset	1050	2271	3300	3420

4   Research Methodology

The following steps are related to the proposed CNN architecture and coarse-to-fine framework as well as the specific used training strategy as in Sects. 4.1 and 4.2 respectively. The last sub-section, Sect. 4.3, shows the evaluation methodology.

4.1   Proposed CNN Architecture

The single-shot-based learning approach is utilised, whereby extracting features of an object, and detecting objects are performed using a single CNN. Effective object detection using CNNs will heavily depend on the meta-architecture a CNN uses for feature extraction. Therefore, the structure/architecture of different state-of-the-art CNN networks were investigated within the wider research context of this for potential use in multi-class object detection being proposed in this paper The study conducted includes how different state-of-the-art architectures differ in terms of the number of convolutional layers, activation function, and type of pooling.

Following the practical evaluation of different state-of-the-art architectures, YOLO-V3 was adopted for the given task. This architecture uses Darknet-53 for the feature extraction, which has 53 convolutional layers, and a further 53 convolutional layers for object detection from the feature map. In total, YOLO-V3 has 106 convolutional layers, with residual blocks. The residual block is the idea inherited from ResNet, which differs significantly from other architectures in that there are no pooling layers in-between the convolutional layers, although a skipping connection is used to reduce the number of parameters. However, while the last layer in YOLO-V3 uses 'average-pooling', in our investigations, 'max-pooling' has proven to reduce the outliers, and subsequently, it has been tuned for evaluation.

4.2   Training Strategies

Training strategy refers to defining a set of parameters to control the training process of a given architecture. The complexity of training deep neural networks results significantly from the sheer number of parameters than can be tuned and the difficultly in predicting the performance in a given application prior to practically configuring and testing such a configuration. This includes the selection of the gradient descent algorithm, batch size, learning rate, optimization method, and number of iterations. However, the crucial parameters that have a significant impact are the learning rate, batch size and the optimization method, and hence this research evaluates their impact on this research dataset. A large batch size, such as 32 or 64, mostly improves the performance compared to a batch size of 2, 4 or 6 even though this is not the case in certain datasets. As it is restricted by the hardware specification, the batch size used in our investigations was fixed at 12 and the data-balancing strategy used ensures the sufficient number of different classes' samples per iteration. However, the learning rate is the most important hyperparameter that can significantly improve the accuracy and speed. A comprehensive explanation of the learning rate and how it affects the sloping toward the minimum loss can be found in [49]. The learning rate has been tuned in YOLO-V3 and changed from 0.001 to 0.0001, omitting the learning rate decay. This means that the weight updates occur more slowly but also consistently in all iterations.

4.3   Evaluation Methodology

The typical evaluation method of learning algorithms is usually based on calculating the precision, recall, and F1-score, as shown in Eqs. 1, 2, and 3. In this paper, these metrics are calculated for each class before the average is taken, which is eventually used to reflect the overall performance. The interpretation/definitions of the terms True Positive (TP) True Negative (TN), False Positive (FP), False Negative (FN), precision and recall are shown in Table 2.

$$precision = \frac{TP}{TP + FP} \tag{1}$$

$$Recall = \frac{TP}{TP + FN} \tag{2}$$

$$F1\ score = 2. \frac{precision \cdot Recall}{Precision + Recall} \tag{3}$$

Table 2. The interpretation of performance evaluation metrics.

Evaluation term	Evaluation interpretation
True positive (TP)	The number of correctly detected objects
False positive (FP)	The number of falsely detected objects
False negative (FN)	The number of missed detections
Precision	The ratio of the correctly detected objects to the total detections; True and False
Recall	The ratio of the correctly detected objects to baseline or ground truth objects in the given dataset

5   Results and Discussion

The experiments are initiated by configuring the dataset as described in Sect. 3. The total number of cropped images (from the large-scale drone image dataset) used in this research dataset is 1050 images. These images have been divided into 85% for training and 15% for testing. As the test set is randomly selected from amongst the 1050 images, the number of bounding boxes that belongs to each of the three classes, differs from image to image. The test set, which contained 157 images, comprised of 173 palm trees, 442 group-of-animals and 374 sheds/animal-shelters. The state-of-the-art CNN architectures, SSD-500 with VGG-16 and ResNet meta-architecture and YOLO-V3, were configured, trained, tested and evaluated, without any changes to their default parameters, except the batch size used, which was set as 12 for YOLO-V3 and set at 4 for SSD. The details of these architectures are shown in Table 3. Based on the initial performance results, YOLO-V3 registered the highest F1-score and it is selected for further optimization.

Table 3. The default parameters of the CNNs architecture in YOLO-V3, SSD 500 with VGG-Net and ResNet architectures (# refers to 'Number of').

CNNs	# Conv layers	Pooling layer	Activation function	Optimization method	Learning rate (LR)	LR decay
SSD-500 VGG-Net	16	Max-pooling	ReLU	Momentum	0.001	0.0005
SSD-500 ResNet	101	Residual blocks Average pooling	ReLU	Momentum	0.0004	0.0005
YOLO-V3	106	Residual blocks Average pooling	Leaky ReLU	Momentum	0.001	0.0005

The SSD-500 with VGG-16 and ResNet meta-architectures and YOLO-V3 have been configured, trained and tested in their ability to detect multi-class objects in drone images. While the former uses 16 convolutional layers for feature extraction, the SSD with ResNet-uses 101 convolutional layers. YOLO-V3 uses 53 layers based on Darknet-53 for feature extraction and a further 53 convolutional layers for detecting objects from the generated feature map. Therefore, these networks have different number of convolution layers in the feature extraction and the object detection phases. The result of multiclass detection in the research shows an F1-score of 0.91 in using YOLO-V3 compared to 0.77 in SSD-500/VGG-Net and 0.83 in SSD-500/ResNet. The influence of the number of convolution layers is clearly shown whereby SSD-500 with VGG-16 registered the lowest F1-score, significantly better as compared with the F1-score of SSD-500 with ResNet. However, YOLO-V3 outperformed both SSD-500 with VGG-16 and with ResNet by a considerable margin as in Table 4. This is because as compared to the five convolutional layers of the detection phase of the SSD architecture, YOLO-V3 has 53 in the detection phase. Further YOLO-V3's feature extraction process is more comprehensive as compared to that if the two SSD based approaches.

Table 4. The result of training three CNNs architectures for drone-based multi-class object detection without any hyper-parameter optimization. (#BX: Number of bounding boxes, TP: True Positive, FN: False Negative, FP: False Positive).

	#BX	TP	FN	FP	Precisions	Recall	F1-score	Avg. confidence
SSD-500 VGG-Net	443	280	163	0	1	0.63	0.77	0.74
SSD-500 Res-Net	443	321	122	0	1	0.72	0.83	**0.78**
YOLO-V3	443	373	70	0	1	0.84	**0.91**	0.77

The precision of the learned model is 1 in detecting all types of objects with the multi-class model generated as there were no 'False Positive (FP)' detections, i.e. objects which are classified as being of a particular type but are not that type. The challenge here are the missed detections of each object type, represented by the 'False Negative (FN)', which is a total of 151 bounding boxes of objects belonging to one of the types of objects not being detected out of a total of 989 possible objects of all types. YOLO-V3 clearly outperforms both SSD-500 based architectures as it clearly shows better results with regards to the performance parameters, recall, F1-score and Average Confidence. Therefore, the use of YOLO-V3 is recommended for multi-class object detection.

With the aim of improving the obtained result of the modest YOLO-V3, the impact of the activation function, pooling method, learning rate and the optimization method are practically evaluated. This helps in investigating the best integration that could influence the performance as in Sects. 5.1–5.4. The optimal selection in each case will be used in the final coarse-to-fine model as in Sect. 5.5.

5.1   The Impact of Different Activation Functions in the Hidden Units

Most CNNs uses either ReLU or Leaky ReLU in the hidden units aiming to activate certain units to pass over the net. While ReLU omits all neurons less than zero, Leaky ReLU allows a small value to present, which has an impact on reducing the number of non-activated neurons. While YOLO-V3 uses Leaky ReLU in its configuration, the impact of changing it to ReLU is practically tested here. This is because the number of classes is still limited, which is three compared to general object detection tasks, which have 80 and more different classes. Reducing the number of activated neurons can resulted in simplifying the model and reducing the overfitting.

In Table 5, the result of changing the activation function is presented. This shows a slight improvement in the FN from 838 out of 989 in the baseline model to 849 based on using the ReLU activated function. Even the F1-score has a slight improvement from 0.91 to 0.92, the combination with tuning different hyper-parameters such as learning rate or pooling layer can give a noticeable improvement.

Table 5. Overall multiclass object detector performance using YOLO-V3 with different activation functions (#BX: Number of bounding boxes).

YOLO-V3	#BX	TP	FN	FP	Precisions	Recall	F1 score	Avg. confidence
Leaky ReLU	989	838	151	0	1	0.84	0.91	0.82
ReLU	989	849	140	0	1	0.85	0.92	0.83

5.2 The Impact of the Pooling Method

The pooling layer can be optionally used in-between the convolutional layers aiming to reducing the number of parameters by taking either the average, max or any other pooling method of a determined receptive field. While ResNet uses the residual blocks, it has no pooling layers in-between the convolutional layers, but it uses the average pooling toward the end of the network. Max pooling is commonly used in modern CNN architectures including VGG-net and AlexNet, and the practitioners claim it has a good reduction of outliers compared to average pooling. A comparison between average pooling and max pooling toward the end of the feature's extraction layers has been evaluated by training the model twice and comparing the performance. The result is shown in Table 6, which reflects a slight improvement when using the max pooling. This is because the model precision is initially 1, which reflects the lower number of outliers in the model. It is suggested to tune the pooling type if the precision of the model is low or if the number of classes where the model tends to lean is higher than in this research case.

Table 6. Overall multiclass object detection performance based on using YOLO-V3 with different pooling type (#BX: Number of bounding boxes).

YOLO-V3	#BX	TP	FN	FP	Precisions	Recall	F1 score	Avg. confidence
Average pooling	989	838	151	0	1	0.84	0.91	0.82
Max-pooling	989	840	149	0	1	0.85	0.92	0.82

5.3   The Impact of Tuning Learning Rate and the Choice of the Learning Rate Decay Method

The learning rate is the most crucial hyperparameter that formulates the training process and the converging time of a given DNN. As it determines the periodic update of the network loss whilst the network is being trained, the higher the value it is set at, less time would be needed to the network loss to converge but will be noisier compared to the use of smaller values. Practically, the learning rate can be tuned between 0.1 and 1. Usually, researchers use lower learning rate values in training on complex datasets. However, a slightly higher value of learning-rate is used if the dataset is easier to train, particularly when using in conjunction with a large volume of data, aiming to reduce the learning time, but with the consequence of the training process to become unstable. As the learning rate substantially affects the training speed of a network, the *learning rate decay*, a parameter that determines the reduction of the learning rate over each epoch (i.e. each iteration of the learning process), can be used to balance between the speeding up the process and converging the network, when the network tends to reach the local minima.

Therefore, given the significant impact of the learning rate on stabilizing the training process, we conducted experiments with the learning rate set to 0.001 and 0.0001. As the research dataset is complex in nature (high intra-class variations in sheds and group-of-animals) and the data availability per class is limited as compared to typical popular object detection datasets, we ignore the time required for the network to train but would consider the stability and convergence of training as crucial. Therefore, the evaluation of the model performance with learning rates of 0.001 and 0.0001, and with the learning rate decay omitted, was conducted. The results are tabulated in Table 7. The results show the significance of performance improvement when the learning rate is 0.0001 and compared to setting it ten times larger, at 0.0001. However, it is noted that the results with a learning rate of 0.001 was obtained at 50,000 iterations, whilst the result with the leaning rate 0.0001 was achieved in 180,000 iterations. The selection of the learning rate for a training task is hence a decision that should be made keeping in mind the complexity of the task to be carried out, time available for training the network and the relative importance of performance metrics such as the precision, recall, F1-score and confidence, which will depend on the application needs.

Table 7. Overall multiclass object detection performance using YOLO-V3 based on different learning rate. (BX: Number of Bounding Boxes).

YOLO-V3	#BX	TP	FN	FP	Precisions	Recall	F1-score	Avg. confidence
LR = 0.001	989	838	151	0	1	0.84	0.91	0.82
LR = 0.0001	989	870	119	0	1	0.87	0.93	**0.87**

5.4   An Evaluation of the Use of Optimization Methods in Minimizing Loss

As Gradient Descent [7] is the method used for minimizing the loss function, different optimization methods have been used beside it to make the model learn fast and accurately, whereby the Momentum [50] is the most popular approach used by the computer vision research community. However, there is a claim by deep learning practitioners that RMSProp [51] and Adam [52] (combines Momentum with RMSProp) optimizers work better in practice. To evaluate the effectiveness of theses optimizers on the proposed multi-class object detector, we conducted an investigation that effectively used Momentum, RMSProp and Adam optimizers. The results obtained are shown in Table 8.

Table 8. Overall multiclass object detection performance based on YOLO-V3 when different optimizers are used (#BX: Number of bounding boxes).

YOLO-V3	#BX	TP	FN	FP	Precisions	Recall	F1-score	Avg. confidence
Momentum	989	838	151	0	1	0.84	0.91	0.82
RMSProp	989	821	168	0	1	0.83	0.90	0.87
ADAM	989	844	145	0	1	0.85	0.92	0.81

The results in Table 8 demonstrate the slight improvement in object detection that is enabled by the use of ADAM, which is a hybrid between the Momentum and RMSProp optimization approaches.

5.5   Overall Performance of the Optimised Multi-class Object Detector

The experiments conducted in Sects. 5.1–5.4 have been conducted using default settings for all parameters, other than the parameter under investigation. These investigations revealed the efficiencies achievable when the right parameter values are selected and used with the dataset under investigation. Based on the research results presented above that highlighted the optimal setting of each parameter, the YOLO-V3 based multi-class object detector was reconfigured, trained, tested and evaluated. This final model has a ReLU activation function for the hidden units, max pooling toward the end of the network, a learning rate of 0.0001 and an Adam optimizer. The results obtained from this customization are presented in Table 9.

The results tabulated in Table 9 show the performance improvement achieved by the optimised network for multi-class object detection. The True Positive (TP) rate has improved from being 838 total objects accurately detected to 932 accurately detected out of a total of 989 objects annotated in images within bounding boxes. However, as the learning rate has been reduced as compared to the learning rate of the network with default parameters, the total number of iterations to achieve such a result is higher, at 180, 000 iterations. It is noted that the precision of the palm trees has slightly decreased from 1 to 0.99, due to a single case of FP, as shown in Fig. 2. A tree which is not a palm

Table 9. The ultimate YOLO-V3 based multi-class object detector performance.

YOLO-V3	#BX	TP	FN	FP	Precisions	Recall	F1 score	Avg. confidence
Palm trees	173	157	16	1	0.99	0.91	0.95	0.87
Cattle	443	420	22	0	1	0.95	0.97	0.84
Sheds	373	355	19	0	1	0.95	0.97	0.91
Overall	**989**	**932**	**40**	**1**	**1**	**0.94**	**0.96**	**0.87**

Fig. 2. The single case of false detection (False Positive, FP) of palm trees in the test set.

tree that has some perceptual similarity to a palm tree, when seen at a low resolution, has been detected as a palm tree.

Examples of multi-class object detection with the optimized YOLO-V3 CNN are illustrated in Fig. 3a, and 3b. It reflects the ability of the learned model to detect different types of sheds oriented in different angles, different group-of-animals (animal) with different spatial-densities and occlusions and different sizes of palm trees in the drone-based footage. It is noted that the missed palm trees are those that are either very small in size or of a very low resolution. The missed sheds are those that are oriented differently to the orientation of the majority of buildings used in training and the missed group-of-animals are those groups that are sparsely spread within the farm.

Figure 4 illustrates further examples where some types of objects are missed. Despite the above missed and false detections, the multi-class object detector developed in this chapter has an improved rate of missed detections (6%) as compared to a 16% missed detections that resulted from the model with non-optimal parameters.

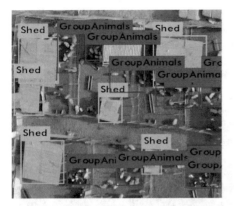

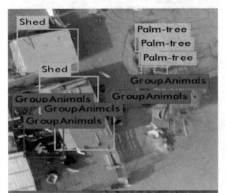
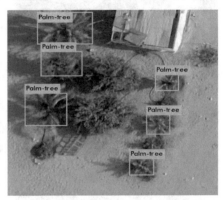

Fig. 3a. The final results of the object detection performance of the proposed methodology for drone-based multi-class object detection

Fig. 3b. The final results of the object detection performance of the proposed methodology for drone-based multi-class object detection

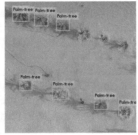

Fig. 4. Examples of missed detections that result from the YOLO-V3 CNN, trained for multi-class object detection

6 Summary and Conclusion

In this paper, multi-class object detection in drone images was investigated, making the best use of the state-of-the-art CNN architectures, SSD-500 supported by the meta-architectures VGG-16 and ResNet and the YOLO-V3 CNN architecture. The key focus of this paper was to develop a single CNN model that is capable of detecting palm-trees, group-of-animals, and sheds/animal-shelters. Initially the performance of all three CNN models were compared in the multi-class object detection task analysing in detail their performance in detecting all three types of objects accurately under default hyper-parameter value selections. This experiment concluded that YOLO-V3 has superior performance to the two SSD-500 based CNN models in recall, F1-score, and average confidence while all three models provided a precision of 1.

Further detailed investigations were subsequently conducted to decide on the optimal hyper-parameter settings when using YOLO-V3 in the given multi-class object detection tasks. Specifically, the impact of using different activation functions, pooling methods, learning rates and optimisation methods to minimize loss were investigated and the relevant optimal parameters were obtained. The original YOLO-V3 network was then reconfigured with these optimal parameters and the model was re-trained, tested and evaluated. The experiment concluded the ability of the optimised YOLO-V3 CNN model to perform significantly better in multi-class object detection in drone images. All performance metrics were substantially improved. Missed detections were carefully studies to make conclusions that due to the high intra-class variations present in all three types of objects, specifically in animal shelters/sheds, significant amount of balanced examples of such objects need to be used in training, to further improve the performance accuracy of the proposed model.

References

1. Lin, T.-Y., et al.: Microsoft coco: common objects in context. In: European Conference on Computer Vision, pp. 740–755 (2014)
2. Everingham, M., Van Gool, L., Williams, C.K., Winn, J., Zisserman, A.: The pascal visual object classes (voc) challenge. Int. J. Comput. Vis. **88**(2), 303–338 (2010)
3. Drone-based dataset for desert area. Falcon Eye Drones Ltd, Dubai, UAE (2017)

4. Aburasain, R.Y., Edirisinghe, E.A., Albatay, A.: Drone-based cattle detection using deep neural networks. In: Arai, K., Kapoor, S., Bhatia, R. (eds.) IntelliSys 2020. AISC, vol. 1250, pp. 598–611. Springer, Cham (2021). https://doi.org/10.1007/978-3-030-55180-3_44

5. Aburasain, R.Y., Edirisinghe, E.A., Albatay, A.: Palm tree detection in drone images using deep convolutional neural networks: investigating the effective use of YOLO V3. In: Conference on Multimedia, Interaction, Design and Innovation, pp. 21–36 (2020)

6. Voulodimos, A., Doulamis, N., Doulamis, Protopapadakis, E.: Deep learning for computer vision: a brief review', Comput. Intell. Neurosci. **2018** (2018)

7. LeCun, Y., Bottou, L., Bengio, Y., Haffner, P.: Gradient-based learning applied to document recognition. Proc. IEEE **86**(11), 2278–2324 (1998)

8. Krizhevsky, A., Sutskever, I., Hinton, G.E.: Imagenet classification with deep convolutional neural networks. In: Pereira, F., Burges, C.J.C., Bottou, L., Weinberger, K.Q. (eds.) Advances in Neural Information Processing Systems, vol. 25, pp. 1097–1105. Curran Associates, Inc. (2012). http://papers.nips.cc/paper/4824-imagenet-classification-with-deep-convolutional-neural-networks.pdf. Accessed 22 Feb 2017

9. Zeiler, M.D., Fergus, R.: Visualizing and understanding convolutional networks. In: European Conference on Computer Vision, pp. 818–833 (2014)

10. Simonyan, K., Zisserman, A.: Very deep convolutional networks for large-scale image recognition. ArXiv Prepr. ArXiv14091556 (2014)

11. He, K., Zhang, X., Ren, S., Sun, J.: Deep residual learning for image recognition. In: Proceedings of the IEEE Conference on Computer Vision And Pattern Recognition, pp. 770–778 (2016)

12. Szegedy, C., Ioffe, S., Vanhoucke, V., Alemi, A.: Inception-v4, Inception-ResNet and the Impact of Residual Connections on Learning (2016). https://arxiv.org/abs/1602.07261v2. Accessed 12 May 2019

13. Howard, A.G., et al.: Mobilenets: Efficient convolutional neural networks for mobile vision applications. ArXiv Prepr. ArXiv170404861 (2017)

14. Fei-Fei, L., Deng, J., Li, K.: ImageNet: Constructing a large-scale image database. J. Vis. **9**(8), 1037 (2009)

15. Szegedy, C., et al.: Going deeper with convolutions. In: Proceedings of the IEEE Conference on Computer Vision and Pattern Recognition, pp. 1–9 (2015)

16. Szegedy, C., Vanhoucke, V., Ioffe, S., Shlens, J., Wojna, Z.: Rethinking the inception architecture for computer vision. In: Proceedings of the IEEE Conference on Computer vision and Pattern Recognition, pp. 2818–2826 (2016)

17. Girshick, R., Donahue, J., Darrell, T., Malik, J.: Rich Feature Hierarchies for Accurate Object Detection and Semantic Segmentation, pp. 580–587 (2014). http://openaccess.thecvf.com/content_cvpr_2014/html/Girshick_Rich_Feature_Hierarchies_2014_CVPR_paper.html. Accessed 17 Jan 2020

18. Girshick, R.: Fast R-CNN, pp. 1440–1448 (2015). http://openaccess.thecvf.com/content_iccv_2015/html/Girshick_Fast_R-CNN_ICCV_2015_paper.html. Accessed 26 Apr 2019

19. Ren, S., He, K., Girshick, R., Sun, J.: Faster R-CNN: towards real-time object detection with region proposal networks. In: Cortes, C., Lawrence, N.D., Lee, D.D., Sugiyama, M., Garnett, R. (eds.) Advances in Neural Information Processing Systems, vol. 28, pp. 91–99. Curran Associates, Inc. (2015). http://papers.nips.cc/paper/5638-faster-r-cnn-towards-real-time-object-detection-with-region-proposal-networks.pdf

20. Liu, W., et al.: Ssd: Single shot multibox detector. In: European Conference on Computer Vision, pp. 21–37 (2016)

21. Redmon, J., Divvala, S., Girshick, R., Farhadi, A.: You only look once: unified, real-time object detection. In: Proceedings of the IEEE Conference on Computer Vision and Pattern Recognition, pp. 779–788 (2016)

22. Uijlings, J.R., Van De Sande, K.E., Gevers, T., Smeulders, A.W.: Selective search for object recognition. Int. J. Comput. Vis. **104**(2), 154–171 (2013)
23. Dai, J., Li, Y., He, K., Sun, J.: R-FCN: Object Detection via Region-based Fully Convolutional Networks (2016). https://arxiv.org/abs/1605.06409v2. Accessed 5 May 2019
24. Redmon, J.: Darknet: Open source neural networks in c (2013)
25. Redmon, J., Farhadi, A.: Yolov3: An incremental improvement ArXiv Prepr. ArXiv180402767 (2018)
26. Radovic, M., Adarkwa, O., Wang, Q.: Object recognition in aerial images using convolutional neural networks. J. Imaging **3**(2), 21 (2017). https://doi.org/10.3390/jimaging3020021
27. Ikshwaku, S., Srinivasan, A., Varghese, A., Gubbi, J.: Railway corridor monitoring using deep drone vision. In: Computational Intelligence: Theories, Applications and Future Directions - Volume II, pp. 361–372 (2019)
28. Al-Sa'd, M.F., Al-Ali, A., Mohamed, A., Khattab, T., Erbad, A.: RF-based drone detection and identification using deep learning approaches: an initiative towards a large open source drone database. Future Gener. Comput. Syst. **100**, 86–97 (2019). https://doi.org/10.1016/j.fut ure.2019.05.007
29. Varghese, A., Gubbi, J., Sharma, H., Balamuralidhar, P.: Power infrastructure monitoring and damage detection using drone captured images. In: 2017 International Joint Conference on Neural Networks (IJCNN), pp. 1681–1687 (2017). https://doi.org/10.1109/IJCNN.2017.796 6053
30. Shao, W., Kawakami, R., Yoshihashi, R., You, S., Kawase, H., Naemura, T.: Cattle detection and counting in UAV images based on convolutional neural networks. Int. J. Remote Sens. **41**(1), 31–52 (2020). https://doi.org/10.1080/01431161.2019.1624858
31. Kellenberger, B., Volpi, M., Tuia, D.: Fast animal detection in UAV images using convolutional neural networks. In: 2017 IEEE International Geoscience and Remote Sensing Symposium (IGARSS), pp. 866–869 (2017)
32. Malek, S., Bazi, Y., Alajlan, N., AlHichri, H., Melgani, F.: Efficient framework for palm tree detection in UAV images. IEEE J. Sel. Top. Appl. Earth Obs. Remote Sens. **7**(12), 4692–4703 (2014). https://doi.org/10.1109/JSTARS.2014.2331425
33. Moreira, A.: Estimating babassu palm density using automatic palm tree detection with very high spatial resolution satellite images - ScienceDirect (2017). https://www.sciencedirect. com/science/article/pii/S0301479717301081. Accessed 27 Jun 2019
34. Wang, Y., Zhu, X., Wu, B.: Automatic detection of individual oil palm trees from UAV images using HOG features and an SVM classifier. Int. J. Remote Sens. **40**(19), 7356–7370 (2019). https://doi.org/10.1080/01431161.2018.1513669
35. Al Mansoori, S., Kunhu, A., Al Ahmad, H.: Automatic palm trees detection from multi-spectral UAV data using normalized difference vegetation index and circular Hough transform. In: High-Performance Computing in Geoscience and Remote Sensing VIII, vol. 10792, p. 1079203 (2018)
36. AlMaazmi, A.: Palm trees detecting and counting from high-resolution WorldView-3 satellite images in United Arab Emirates. In: Remote Sensing for Agriculture, Ecosystems, and Hydrology XX, vol. 10783, p. 107831M (2018). https://doi.org/10.1117/12.2325733
37. Freudenberg, M., Nölke, N., Agostini, A., Urban, K., Wörgötter, F., Kleinn, C.: Large scale palm tree detection in high resolution satellite images using U-Net. Remote Sens. **11**(3), 312 (2019). https://doi.org/10.3390/rs11030312
38. Mubin, N.A., Nadarajoo, E., Shafri, H.Z.M., Hamedianfar, A.: Young and mature oil palm tree detection and counting using convolutional neural network deep learning method. Int. J. Remote Sens. **40**(19), 7500–7515 (2019). https://doi.org/10.1080/01431161.2019.1569282
39. Zortea, M., Nery, M., Ruga, B., Carvalho, L.B., Bastos, A.C.: Oil-palm tree detection in aerial images combining deep learning classifiers. In: IGARSS 2018 - 2018 IEEE International

Geoscience and Remote Sensing Symposium, pp. 657–660 (2018). https://doi.org/10.1109/IGARSS.2018.8519239

40. Yousif, H., Yuan, J., Kays, R., He, Z.: Fast human-animal detection from highly cluttered camera-trap images using joint background modeling and deep learning classification. In: 2017 IEEE International Symposium on Circuits and Systems (ISCAS), pp. 1–4 (2017). https://doi.org/10.1109/ISCAS.2017.8050762

41. Gomez Villa, A., Salazar, A., Vargas, F.: Towards automatic wild animal monitoring: Identification of animal species in camera-trap images using very deep convolutional neural networks. Ecol. Inform. **41**, 24–32 (2017). https://doi.org/10.1016/j.ecoinf.2017.07.004

42. Norouzzadeh, M.S., et al.: Automatically identifying, counting, and describing wild animals in camera-trap images with deep learning. Proc. Natl. Acad. Sci. **115**(25), E5716–E5725 (2018). https://doi.org/10.1073/pnas.1719367115

43. Rivas, A., Chamoso, P., González-Briones, A., Corchado, J.M.: Detection of cattle using drones and convolutional neural networks. Sensors **18**(7), 2048 (2018). https://doi.org/10.3390/s18072048

44. Saqib, M., Khan, S.D., Sharma, N., Blumenstein, M.: A study on detecting drones using deep convolutional neural networks. In: 2017 14th IEEE International Conference on Advanced Video and Signal Based Surveillance (AVSS), pp. 1–5 (2017). https://doi.org/10.1109/AVSS.2017.8078541

45. Kim, B.K., Kang, H.-S., Park, S.-O.: Drone classification using convolutional neural networks with merged doppler images. IEEE Geosci. Remote Sens. Lett. **14**(1), 38–42 (2016)

46. Von Luxburg, U.: A tutorial on spectral clustering. Stat. Comput. **17**(4), 395–416 (2007)

47. Buettner, R., Baumgartl, H.: A Highly Effective Deep Learning Based Escape Route Recognition Module for Autonomous Robots in Crisis and Emergency Situations (2019). http://scholarspace.manoa.hawaii.edu/handle/10125/59506. Accessed 05 Jun 2019

48. Yoon, I., Anwar, A., Rakshit, T., Raychowdhury, A.: Transfer and online reinforcement learning in STT-MRAM based embedded systems for autonomous drones. In: 2019 Design, Automation Test in Europe Conference Exhibition (DATE), pp. 1489–1494 (2019). https://doi.org/10.23919/DATE.2019.8715066

49. Aburasain, R.Y.: Application of convolutional neural networks in object detection, re-identification and recognition. Loughborough University (2020)

50. Polyak, B.T.: Some methods of speeding up the convergence of iteration methods. USSR Comput. Math. Math. Phys. **4**(5), 1–17 (1964)

51. Kurbiel, T., Khaleghian, S.: Training of Deep Neural Networks based on Distance Measures using RMSProp. ArXiv170801911 Cs Stat (2017). http://arxiv.org/abs/1708.01911. Accessed 15 Apr 2020

52. Kingma, D.P., Ba, J.: Adam: A Method for Stochastic Optimization. ArXiv14126980 Cs (2017). http://arxiv.org/abs/1412.6980. Accessed 15 Apr 2020

Sentiment Analysis Using State of the Art Machine Learning Techniques

Salih Balci[1], Gozde Merve Demirci[2], Hilmi Demirhan[3], and Salih Sarp[4(✉)]

[1] Gebze Technical University, Kocaeli, Turkey
s.balci2020@gtu.edu.tr
[2] The Graduate Center CUNY, New York, USA
gdemirci1@gradcenter.cuny.edu
[3] University of North Carolina, Chapel Hill, USA
hd1526@uncw.edu
[4] Virginia Commonwealth University, Richmond, USA
sarps@vcu.edu

Abstract. Sentiment analysis is one of the essential and challenging tasks in the Artificial Intelligence field due to the complexity of the languages. Models that use rule-based and machine learning-based techniques have become popular. However, existing models have been under-performing in classifying irony, sarcasm, and subjectivity in the text. In this paper, we aim to deploy and evaluate the performances of the State-of-the-Art machine learning sentiment analysis techniques on a public IMDB dataset. The dataset includes many samples of irony and sarcasm. Long-short term memory (LSTM), bag of tricks (BoT), convolutional neural networks (CNN), and transformer-based models are developed and evaluated. In addition, we have examined the effect of hyper-parameters on the accuracy of the models.

Keywords: Sentiment analysis · Bag of tricks · Transformer · BERT · CNN

1 Introduction

Over a decade, so many users have generated lots of content on the internet, mostly on social platforms. Millions of individuals use any social platform to express their opinions, and these contents create a considerable amount of raw data. Such a massive volume of raw data brings lots of exciting tasks. These tasks are under Natural Language Processing (NLP) applications. NLP is a branch of Artificial Intelligence (AI) focusing on text-related problems, and one of its goals is to understand the human language. In NLP, there are various language-related fields to focus on, such as machine translation, chatbots, summarizations, question answering, sentiment analysis [1, 2].

Sentiment Analysis (SA) is closely linked to NLP. Sentiment analysis is a scale result that shows the sentiment and the opinions coming from a raw text. It is essential and helpful application to analyze an individual's thoughts. The sentiment analysis result may help various fields from industrial purposes such as advertising and sales to academic

C. Biele et al. (Eds.): MIDI 2021, LNNS 440, pp. 34–42, 2022.
https://doi.org/10.1007/978-3-031-11432-8_3

purposes. Even though sentiment analysis has been a focus of authors for a while, the challenges in this field, such as having sarcasm and irony in a text, make this task still unfinished. Therefore, there is still colossal attention on sentiment analysis, and new approaches have arisen [3].

Recently, many novel approaches to AI systems have been developed using Machine Learning (ML). Also, with the help of Deep Learning (DL) techniques, a subfield of ML, the algorithms such as Generative Adversarial Networks and transformers improved the performance of AI tasks significantly [4]. Many studies have focused on sentiment analysis in NLP fields [5–7]. Today, comprehensive survey studies and novelty approaches to sentiment analysis are still being carried out. In [5], the authors created an extensive research survey on sentiment analysis. In the paper, levels of sentiment analysis, challenges and trends in this field, and the genetic process are mentioned detail. Here, sarcasm detection was shown as one of the challenges, and related studies to solve this challenge are examined. Instead of traditional machine learning approaches, other techniques such as DL and reinforcement learning resulted in more robust solutions to challenges.

Authors in [6] proposed a hybrid model by combining the DL approach and sentiment analysis model to predict the stock prices. Sentiment analysis in the stock market is critical to estimating future price changes. In this article, the authors created a hybrid model using a Convolutional Neural Network (CNN) to create a sentiment analysis classifier on investors' comments and Long Short-Term Memory (LSTM) Neural Network to analyze the stock. Implementation of this hybrid model on the real-life data on the Shanghai Stock Exchange (SSE) showed that the hybrid approach outperformed.

In the paper [7], the study conducts a novel approach to ML-based classifiers. From Twitter, related tweets have been retrieved from eight countries, and people's behavior on the infectious disease was aimed to analyze. In the proposed model, Naïve Bayes Support Vector Machines (NBSVM), CNN, Bidirectional Gated Recurrent Network (BiGRU), fastText, and DistilBERT [8] were used as base classifiers, and the fusion of these approaches was represented as "Meta Classifier". The proposed model gave better results than four DL and one machine learning approach.

This paper gives the comparison works on sentiment analysis using state-of-art ML approaches: LSTM, Bag of Words (BoT), CNN, and transformer. The aim is to compare the performances of deep learning approaches in terms of accuracy and time complexity. Moreover, the impact of hyperparameters on the model's accuracy was analyzed.

This paper is organized as follows. In the second section, background information about sentiment analysis, particularly in NLP, is discussed. In the third section, the approach to the methodology and the state-of-art approach are explained. The fourth section explains the results, and the paper is concluded in the last section.

2 Background Information

2.1 Preprocessing the Text Data

The abundance of text data provides many opportunities for training the NLP models. However, the unstructured nature of the text data requires preprocessing. Lowercasing, spelling corrections, punctuation, and stop word removal are some of these preprocessing steps. These operations could be easily implemented in Python language using NumPy, pandas, textblob, or nltk libraries. Then, tokenization, stemming, and lemmatization processes are realized to convert raw text data to smaller units with removing redundancy. The tokenization process splits the stream of text into words [9]. Extracting the root of a word is done using stemming techniques. Similar to stemming, the lemmatizing process extracts the base form of a word. These preprocessing steps could be implemented using textblob or nltk libraries as well. The data used in this study are taken from the public IMDB dataset. It has binary labeled 50000 reviews. To train the SA models, 17500 reviews are chosen for training, 7500 for validation, and 25000 for testing purposes.

2.2 Feature Engineering

After preprocessing, feature extraction steps are implemented to transform words or characters to computer understandable format. This step includes vectorization. Vectorization of words provides corresponding vectors for further processing. Relating the representations of the words with similar meanings is achieved using word embedding. Each word is represented with a vector using feature engineering methods such as N-grams, count vectorizing, and term frequency-inverse document frequency (TF-IDF). Word embedding methods are developed to capture the semantic relation between the words [10]. Word2vec and fastText frameworks are developed to train word embeddings. Skip-Gram and Continuous Bag of Words (CBOW) are commonly used models [11]. This study benefits from the pre-trained GloVe word embeddings to increase the performance of the sentiment analysis models.

3   Methodology

This section develops four major sentiment analysis models using ML techniques, i.e., LSTM, BoT, CNN, and transformer.

3.1   LSTM-Based Sentiment Analysis

Neural networks utilize backpropagation algorithms to update the weights using chain rule in calculus. For large deep neural networks, backpropagation could cause troubles such as vanishing or exploding gradients. Long-term memory architecture is an improved version of the Recurrent Neural Network (RNN) to overcome the vanishing gradient problem with an extra recurrent state called a memory cell. LSTM achieves long-range data series learning, making it a suitable technique for a sentiment analysis task.

Forward and backward RNNs are combined to form a single tensor to increase the performance of the LSTM-based model. In addition to bidirectionality, multiple LSTM layers could be stacked on top of each other to increase the performance further. We have used two layers of LSTM and 0.5 dropout on hidden states for regularization to decrease the probability of overfitting.

3.2 Bag of Tricks-Based Sentiment Analysis

Among DL techniques, linear classifiers could achieve similar performances with a more straightforward design [12]. However, one of the disadvantages of linear classifiers is their inability to share the parameters among features and classes [12]. The Bag of Tricks (BoT) architecture uses linear models with a rank constraint and a fast loss approximation as a base. Means of words' vector representations are fed to linear classifiers to get a probability distribution of a class. In this study, bag of n-grams, i.e., bigrams, are utilized instead of word order for higher performance where the n-gram technique stores n-adjacent words together.

This architecture also does not use pre-trained word embeddings, which could ease its usage in other languages that do not yet have efficient pre-trained word embeddings. This model has fewer parameters than the other models, and the results are comparable in less time.

3.3 CNN (Convolutional Sentiment Analysis)-Based Sentiment Analysis

Convolutional Neural Network (CNN) is a DL approach used primarily for raw data. CNN has a wide scope of application fields from image recognition to NLP with its architecture [13, 18, 19]. CNN has a multi-layered feed-forward Neural Network architecture. It aims to reduce the data into a shape so that it does not lose any features while processing. This way, CNN makes sure that the prediction accuracy and quality will be higher. There are convolutional and pooling layers in its architecture to reshape the data while training. Traditionally, CNNs have one or more convolutional layers followed by one or more linear layers.

In the convolutional layer, the data is processed and reshaped by kxk filters, usually k = 3. Each filter has a form in the architecture, and this filter gives the weight for the data points. The intuitive idea behind learning the weights is that the convolutional layers act as feature extractors, extracting parts of the most critical data. This way, the dominant features are extracted.

CNN has been mainly used for image fields for a long time; image recognition, image detection, analysis, etc. However, it is started to be used in NLP approaches, and it gives significant results. In this study, convolutional layers will be used as "k" consecutive words in a piece of text. The kxk filter mentioned in the convolutional layer would represent a patch of an image in image related field. However, here 1xk filter will be used to focus on k consecutive words as in bi-grams (a 1x2 filter), tri-grams (a 1x3 filter), and/or n-grams (a 1xn filter) inside the text.

3.4 Transformer Based Sentiment Analysis

The transformer is a state of art network architecture proposed in 2017 [14]. In this state-of-art approach, the results showed that with the use of a transformer, NLP tasks outperformed other techniques. After transformer architecture, various models focusing on NLP fields such as ROBERT [15], BERT [16], ELECTRA [17] were proposed. Specifically, BERT (Bidirectional Encoder Representations from Transformers) model is one of the most robust state-of-art approaches on NLP fields. BERT was introduced in 2019 by Google AI Language, and since then, it has started to be used very quickly in academics and industry. BERT is a pre-trained model which is very easy to fine-tune model into our dataset. It has a wide range of language options [16].

BERT architecture is a multi-layer bidirectional transformer encoder. BERT input representation has three embedding layers: position embedding, segment embedding, and token embedding. In the pre-training part, BERT applied two unsupervised tasks, Masked LM (MLM) and Next Sentence Prediction (NSP), instead of traditional sequence modeling. BERT has pre-trained with more than 3,000 M words.

In the study, we used the transformers library to obtain a pre-trained BERT model employed as our embedding layers. We only trained the rest of the model with the pre-trained architecture itself, which learns from the transformer's representations. The transformer provides a pooled output and the embedding for the whole sequence. Due to the purpose of this study (sentiment outputs), the model has not utilized the pooled output from the architecture.

The input sequence of the data was tokenized and trimmed to the maximum sequence size. The tokenized input was converted to a tensor and prepared for fine-tuning. After fine-tuning the model, it was then used to evaluate the sentiment of various sequences.

4   Results and Discussion

This section compares performances of state-of-the-art (SOTA) models in terms of accuracy, time, and loss.

4.1   Time Analysis

The training time comparisons of SOTA models are indicated in Table 1. The results indicated that most DL models provide reasonable training time except the transformer-based model. Models that use LSTM, BoT, and CNN performed an epoch per minute, whereas the BoT-based model achieves 13 s per epoch in contrast to 28 min in the case of the transformer model. In the testing phase, results are aligned with the training phase. Even though only time analysis does not give a concrete interpretation of a model, we see a considerable time efficiency difference between BERT and other models.

Table 1. Training and testing time comparison of SOTA models

Epoch/Test	LSTM	BoT	CNN	Transformer
1	1 m 41 s	0 m 14 s	0 m 30 s	28 m 4 s
2	1 m 40 s	0 m 13 s	0 m 30 s	28 m 7 s
3	1 m 41 s	0 m 13 s	0 m 30 s	28 m 6 s
4	1 m 40 s	0 m 13 s	0 m 30 s	28 m 7 s
5	1 m 40 s	0 m 13 s	0 m 30 s	28 m 7 s

(continued)

Table 1. (*continued*)

Epoch/Test	LSTM	BoT	CNN	Transformer
6	1 m 41 s	0 m 13 s	0 m 30 s	28 m 6 s
7	1 m 41 s	0 m 13 s	0 m 30 s	28 m 4 s
8	1 m 40 s	0 m 13 s	0 m 30 s	28 m 7 s
9	1 m 40 s	0 m 13 s	0 m 30 s	27 m 56 s
10	1 m 40 s	0 m 13 s	0 m 30 s	27 m 58 s
Testing	15 ms	9 ms	10 ms	35 ms

4.2   Validation and Test Losses

Validation loss is another critical metric to evaluate how a model fits new data. Validation loss is also a good indicator of overfitting. The models' validation, training, and test losses are shown in Fig. 1 and Table 2.

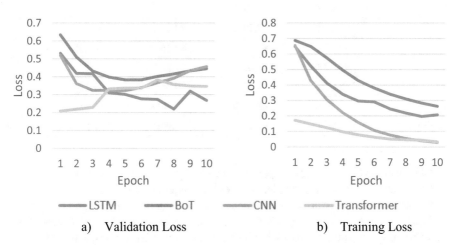

a)   Validation Loss   b)   Training Loss

Fig. 1. Validation and training losses of the models.

The loss graph of the transformer-based model indicates that it could converge faster than other models with fewer training epochs. This will be a result of pre-training of the transformer model.

Table 2. Test losses of the models.

Models	LSTM	BoT	CNN	Transformer
Loss	0.323	0.391	0.344	0.209

4.3   Validation Accuracy

Validation accuracy in combination with validation loss could be used to determine the model's generalization ability. The validation and testing accuracies of the models are given in Table 3. Validation accuracy reveals that five epochs of training are enough to get good results which also in line with the validation loss. Testing accuracy is aligned with the validation accuracy where the transformer-based model achieves the best performance.

Table 3. Validation and testing accuracies of the models.

Epoch/Test	LSTM	BoT	CNN	Transformer
1	73.28%	72.08%	77.03%	91.93%
2	82.19%	76.29%	84.56%	91.76%
3	79.64%	80.50%	86.06%	92.02%
4	87.71%	83.55%	86.78%	90.74%
5	87.81%	85.47%	86.99%	91.31%
6	89.27%	85.47%	87.23%	91.31%
7	89.65%	87.09%	87.16%	90.89%
8	91.52%	87.68%	87.30%	91.19%
9	88.06%	88.07%	86.96%	92.15%
10	89.69%	88.46%	87.40%	91.85%
Testing	86.96%	85.18%	85.04%	91.58%

Observations derived from the performance comparisons are outlined below.

Observation 1: BoT-based model is faster than other DL models.

Observation 2: Transformer-based model takes a long time to train and predict.

Observation 3: Optimum epoch number could be determined using accuracy and loss of training and validation phases. Five epochs of training provide optimum training.

Observation 4: Transformer-based model converges faster than other models.

5   Conclusion

Sentiment analysis has been studied to harness the reviews, comments, and other written documents. The potential of sentiment analysis provided many benefits to various industries such as entertainment and e-commerce. This paper presents sentiment analysis models that utilize four ML techniques, i.e., LSTM, BoT, CNN, and transformer. Their performances in terms of time, loss, and accuracy are examined and compared. The BoT-based sentiment analysis model is faster than other ML models, whereas the transformer-based model performs poorly in terms of time. Furthermore, this study also demonstrates the accuracies of these models. The transformer-based sentiment analysis model achieved higher accuracy than other ML models.

This study indicates that ML techniques could be utilized successfully for sentiment analysis tasks. It is expected that this study will be helpful for both developers and researchers while deploying ML-based sentiment analysis algorithms into their projects.

References

1. Balki, F., Demirhan, H, Sarp, S.: Neural machine translation for Turkish to English using deep learning. In: Biele, C., Kacprzyk, J., Owsiński, J.W., Romanowski, A., Sikorski, M. (eds.) Conference on Multimedia, Interaction, Design and Innovation. AISC, vol. 1376, pp. 3–9. Springer, Cham (2020). https://doi.org/10.1007/978-3-030-74728-2_1
2. Demirci, G.M., Keskin, Ş.R., Doğan, G.: Sentiment analysis in Turkish with deep learning. In: 2019 IEEE International Conference on Big Data (Big Data), pp. 2215–2221. IEEE (2019)
3. Farzindar, A., Inkpen, D.: natural language processing for social media. Syn. Lect. Hum. Lang. Technol. **8** 1–166 (2015)
4. Sarp, S., Kuzlu, M., Wilson, E., Guler, O.: WG2AN: synthetic wound image generation using generative adversarial network. J. Eng. **2021**(5), 286–294 (2021)
5. Birjali, M., Kasri, M., Beni-Hssne, A.: A comprehensive survey on sentiment analysis: approaches, challenges and trends. Knowl. Based Syst. **226**, 107134 (2020)
6. Jing, N., Wu, Z., Wang, H.: A hybrid model integrating deep learning with investor sentiment analysis for stock price prediction. Expert Syst. Appl. **178**, 115019 (2021)
7. Basiri M.E., Nemat, S., Abdar, M., Asadi, S., Acharrya, U.R.: A novel fusion-based deep learning model for sentiment analysis of COVID-19 tweets. Knowl. Based Syst. **228** (2020)
8. Sanh, V., Debut, L., Chaumond, J., Wolf, T.: DistilBERT, a distilled version of BERT: smaller, faster, cheaper and lighter. arXiv preprint arXiv:1910.01108 (2019)
9. Vijayarani, S., Janani, R.: Text mining: open source tokenization tools-an analysis. Adv. Comput. Intell. Int. J. **3**(1), 37–47 (2016)
10. Kulkarni, A., Shivananda, A.: Natural language processing recipes. Apress (2019)
11. Mikolov, T., Chen, K., Corrado, G., Dean, J.: Efficient estimation of word representations in vector space. arXiv preprint arXiv:1301.3781 (2013)
12. Joulin, A., Grave, E., Bojanowski, P., Mikolov, T.: Bag of tricks for efficient text classification. arXiv preprint arXiv:1607.01759 (2016)
13. Sarp, S., Kuzlu, M., Pipattanasomporn, M., Guler, O.: Simultaneous wound border segmentation and tissue classification using a conditional generative adversarial network. J. Eng. **2021**(3), 125–134 (2021)
14. Vaswani, A., et al.: Attention is all you need. In: Advances in Neural Information Processing Systems, pp. 5998–6008 (2017)
15. Pappagari, R., Zelasko, P., Villalba, J., Carmiel, Y., Dehak, N.: December. Hierarchical transformers for long document classification. In: 2019 IEEE Automatic Speech Recognition and Understanding Workshop (ASRU), pp. 838–844. IEEE (2019)
16. Devlin J., Chang, M., Lee, K., Toutanova, K.: BERT: Pre-training of Deep Bidirectional Transformers for Language Understanding. NAACL (2019)
17. Clark, K., Luong, M.T., Le, Q.V., Manning, C.D.: Electra: Pre-training text encoders as discriminators rather than generators. arXiv preprint arXiv:2003.10555 (2020)
18. Sarp, S., Kuzlu, M., Wilson, E., Cali, U., Guler, O.: The enlightening role of explainable artificial intelligence in chronic wound classification. Electronics **10**(12), 1406 (2021)
19. Yin, W., Kann, K., Yu, M., Schütze, H.: Comparative study of CNN and RNN for natural language processing. arXiv preprint arXiv:1702.01923 (2017)

Causal Inference - Time Series

Aishwarya Asesh$^{(\boxtimes)}$

Adobe, Mountain View, USA
a.asesh@gmail.com

Abstract. Detecting causation in observational data is a difficult task. Identifying the causative direction, coupling delay, and causal chain linkages from time series may be used to find causal relationships. Three issues must be addressed when inferring causality from time series data: resilience to noisy time series, computing efficiency and seamless causal inference from high-dimensional data. The research aims to provide empirical evidence on the relationship of *Marvel Cinematic Universe (MCU)* movies and marvel comic book sales using Fourier Transforms and *cross-correlation* of two time series data. The first of its kind study, establishes some concrete evidence on whether the trend of declining comic study and increasing movie audience will disrupt in the post COVID world.

Keywords: Time series · Causality · Causal inference · Correlation

1 Introduction

Movies based on comic books have become some of Hollywood's most popular in the previous two decades. However, due in part to a speculative bubble that resulted in a crash[1] and in part to the development of digital media, comic book sales have fallen from their high levels in the 1990s.s. The huge success of the super-hero genre on the big screen should, in theory, translate to more interest in the comic book source material. This study is focused on determining the extent to which this link exists.

Fourier Transforms and *cross-correlation* of two time series are used in this study to evaluate and comprehend the total monthly profits of comic books and comic book movies. The focus is on films from the *Marvel Cinematic Universe (MCU)*, which includes films starring comic book characters such as *Iron Man, Thor, Captain America, and Spider-Man.* These projects contain fairly accurate adaptations of comic-book characters and plots, as well as represent some of the most profitable and successful comic-book films ever, with global box-office receipts of over 22 billion dollars (USD) [1]. Now that Marvel and DC are regularly topping the box office with their superhero films, they deserve to be recognized. They have constantly released more and more superhero movies as

---

[1] https://www.washingtonexaminer.com/weekly-standard/the-crash-of-1993.

C. Biele et al. (Eds.): MIDI 2021, LNNS 440, pp. 43–51, 2022.
https://doi.org/10.1007/978-3-031-11432-8_4

a result of building a cinematic universe, bringing in an increasing number of people to the theaters to view their flicks [2,3].

The study is led by the following hypothetical question: To what degree do MCU film releases impact Marvel comic book sales? On one hand, one can anticipate that, on a wide scale, the frequency of Marvel adaptations throughout the MCU's active period should result in a general increasing trend in Marvel comic-book sales. Variations in comic-book sales around the general trends, on the other hand, should be linked to the release of MCU films [4]. The above question, in terms of time series, is essentially about the similarity of two signals. However, because correlation does not indicate causality, the findings (if any) will be limited to the *coherence* of the hypotheses with the data, which one may or may not be able to confirm or refute.

One of the major incentives for social scientists to use data to detect behavior patterns and uncover intriguing correlations is the quest for causation [5]. The identification of causal effects, also known as causal inference, aims to identify the underlying mechanisms that cause changes in a phenomena [6,7]. While the availability of large data and high-performance computers allows for innovative data analysis using causal inference, only a few research in the field of regional studies use causal inference methodologies. Understanding regional or country-wide phenomena with these advancements, on the other hand, offers vital insights into society and helps us to better monitor global trends [8–10]. What have been the highs and lows in MCU Movies and Marvel Comics sales in the United States over the last two decades? One may not know the causes of these anomalies, but by understanding the causes of phenomena that may have gone unnoticed previously, scientists may be able to better predict the consequences of one pattern change, provide solutions for impending problems, or be prepared for the coming paradigm shift.

2   Data

This project takes data from two sources. For the comic book sales figures, the data is taken from the universal repository of Comichron[2], an internet database featuring monthly comic book sales figures. For the movie earnings the data is collected from Universal Machine Learning (ML) repository. For all time domestic movie sales, Box Office Mojo[3], an internet database giving various data related to movies, including box-office earnings over time, is used.

The two time series under consideration are the *total profits* of all Marvel comics and the *total earnings* of all MCU (domestically: USA) movies, in other words, the overall connections between Marvel movies and Marvel comics. The rationale behind this decision is because individual comic book series sales (for example, The Amazing Spider-Man) are unpredictable and fluctuate depending on a variety of factors, including the writer and artist. On the other hand, one would anticipate publisher-wide sales numbers to be less influenced by external

---

[2] https://www.comichron.com.

[3] https://data.world/eliasdabbas/boxofficemojo-alltime-domestic-data.

factors. The data sets span the months of January 2008 through December 2019, nearly book ending the MCU till date.

Figure 1 displays the total profits of Marvel Comics during the MCU's history (including the years 2008 and 2019).

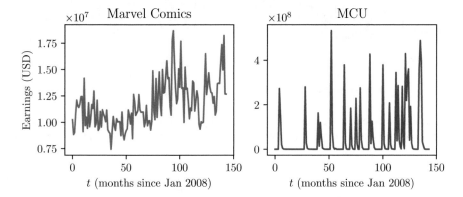

Fig. 1. Marvel Comics profits as a time series t in months since Jan 2008. *Left panel* - Marvel Comics profits as a time series t. *Right panel* - Total Marvel Cinematic Universe domestic box office profits as a function of time t

The MCU dataset (purple trace, right panel) displays a continuous zero signal punctuated by strong peaks coinciding with movie releases, as one might anticipate. The comics data, on the other hand, displays a wide, complex pattern with high-frequency oscillations overlaid. The first step will be to filter the data in order to determine whether or not there is a trend.

3   Experiments

3.1   Filtering

Filtering the data is done to make the trend more obvious. First, a least-squares fit to a degree 5 polynomial is used to estimate the trend. The predicted trend is then subtracted, leaving just the variations around it. Figure 2 shows the raw data, trendline, and detrended data.

The data is filtered using the *Fourier Transform* of the time series f, which is defined as:

$$F_k = \Delta t \sum_{j=0}^{N-1} f_j e^{-i2\pi jk}, \tag{1}$$

in which N is the length of the time series. The Fourier Transform can be seen in Fig. 3.

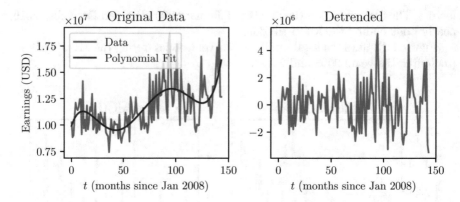

Fig. 2. Raw comic book revenues and a degree 5 polynomial fit - *Left panel*. Subtracting the polynomial trendline from the raw data to get 'detrended' time series - *Right panel*

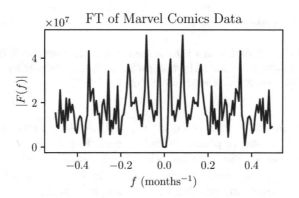

Fig. 3. As a function of frequency f, the amplitude of the fourier transform of the Marvel Comics Earnings time series.

The greatest amplitude peaks have frequencies lower than 0.1 month^{-1} in absolute value, as determined by the amplitude of the peaks, which approximately corresponds to the 'component' of the signal with the appropriate frequency. As a result, any Fourier components with frequencies larger than that value are converted to zero to smooth out the data, thus truncating the Fourier Transform. The smoother trace displayed in Fig. 4 is obtained by doing an inverse Fourier Transform and putting back in the polynomial trend.

There is a general increasing trend over the duration of roughly ten-year frame of investigation, as one would predict based on the polynomial fit (and from eyeballing the raw data). The rise appears to be related to the number of MCU releases, but rather than going into detail about the quantitative elements, this research focuses on the correlation analysis.

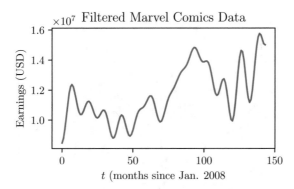

Fig. 4. The result of truncating Fourier Transform and filtering Marvel Comics data

3.2   Correlation

Now moving to examining fluctuations around the general trend after considering the broad effects of rising MCU releases on Marvel comics data. The *cross-correlation* between time series f and g is used to understand how these oscillations connect to the MCU signal:

$$C_{fg}(\tau) = \int_{-\infty}^{\infty} d\tau f^*(t) g(t + \tau). \tag{2}$$

When g is shifted by an amount τ, the above connection in Eq. (2) essentially yields the correlation between the two time series. When C_{fg} is graphed as a function of τ, the peaks indicate the extent to which g is connected to f when g is shifted by τ.

For MCU earnings, the same approach of removing the data as illustrated in Fig. 2 and normalizing the resultant detrended data by dividing by the maximum (absolute value) entry is used to make the two series more equal in magnitude for more appropriate cross-correlation values. Cross-correlation is calculated for the two normalized datasets. The cross-correlation between the MCU and Marvel Comics profits time series is shown in Fig. 5. The $\tau-$axis is set up so that a positive value of τ correlates to a shift of τ in the comics time series (in the positive direction).

4   Results and Discussion

4.1   Results

In terms of answering the guiding questions, findings have certain consequences. For one thing, the pattern seen in Fig. 4 indicates a definite rise with time, which is at least consistent with the predictions. But, as one can see in the previous section, the clear increasing tendency shown by the filtering process drives researchers to investigate the fluctuations that surround it. However, the

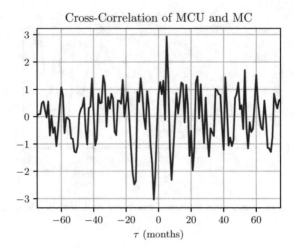

Fig. 5. The MCU earnings dataset and the Marvel Comics (MC) dataset cross-correlate as a function of τ, the time series shift. Peaks define a high-correlation zone.

results of the correlation (Fig. 5) are sloppy and unsatisfactory in various aspects. For example, at $\tau \approx 6$ months, it is observed that one peak jumps out. This peak implies that the comic book data from 6 months ago is connected to the MCU data, which makes sense because the advertising campaign for big-budget movies like those in the MCU would begin around this time, with the publication of numerous 'trailers' and television ads. As a result, findings might imply that the start of this process has a greater impact on comic book sales than the actual release of the movie. There is also a peak around $\tau = 0$, which corresponds to the film's release month, but it blends in with the plot's background noise.

Of course, correlation does not indicate causation. At the very least, the correlation data provided above is *consistent* with the hypothesis that MCU films influence Marvel Comics sales.

4.2   Control

It helps to have a 'control' dataset to make assertions more believable. If the theory is true, sales statistics for publishers other than Marvel at $\tau \approx 6$ months should not show the same pattern. While DC Comics is a more apparent choice, it would not be a good control because its comics are in the same genre as Marvel's. The results of the same analysis performed on the monthly sales statistics for IDW are presented in Fig. 6.

Even the peak at $\tau \approx 5$ has comparable noise, and there are even several stronger peaks near $\tau \approx -55$. As a result, it can be asserted that the correlation peak mentioned in Correlation Section is 'real': it accurately represents the impact of comic book movie advertising on comic book sales.

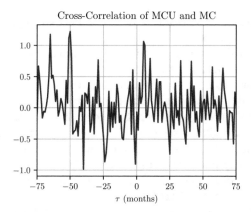

Fig. 6. Cross-correlation of MCU earnings and IDW Comics earnings dataset as a function of τ, the shift between the time series. Peaks give region of high correlation.

4.3 Future Directions

The analyses provided in the preceding sections include a number of aspects that are planned to be carried out in future research. To begin with, noise in data makes it difficult to separate the peaks that are found in the cross-correlation from the background. While the study of the 'control' dataset in the preceding section strengthens the assertions, the connection remains weak.

Second, the fact that the time axis is only accurate to the month level creates a resolution issue. Higher-resolution (daily) comics revenue statistics might reveal more compelling patterns, as it's possible that a surge in comic book sales lasts shorter than a month, causing it to be lost in the monthly graphs. Thus, more granularity of time series data need to be considered for future use cases.

Of course, the amount of variables that impact comic book statistics is another big source of inaccuracy, this study ignores issues like seasonal oscillations (summer is a high-earning month for comic books), inflation, and so on. These factors may probably be modeled in a more complex study; the yearly oscillation, in particular, might be eliminated using a notch filter.

5 Conclusion

The findings in this study do not conclusively prove or disprove the hypothesis, nor do they provide a persuasive response to the central issue concerning the impact of comic book movies on comic book sales. Nonetheless, it is discovered that some pieces of credible evidence would undoubtedly spur further research into the subject. First, it is discovered that filtering monthly comic book sales data indicated a significant visual relationship between the wide growth in comic book sales throughout the ten-year research window and the release of MCU films during the same time. The cross-correlation results imply that the build-up to the release of MCU films (perhaps advertising campaigns) might be quantitatively connected to comic book sales.

The findings of this research illuminates numerous project possibilities, with two primary paths standing out. First, instead of focusing just on Marvel, it would be fascinating to conduct a similar research utilizing *all* comic films vs *all* super-hero comic books, one could discover an even greater link. Second, one might take the reverse approach and compare the sales of certain comic book series to specific film franchises (for example, Avengers comic book vs. Avengers film series). The link would most likely be weak for popular comic book series (readership may 'saturate' at certain point). However, movies can be anticipated to boost interest in the comics for minor characters who are granted movie franchises (*e.g.* Doctor Strange, perhaps).

Overall, this research demonstrates a creative use of the Fourier, filtering, and correlation techniques. The impact of big-budget blockbusters on source material is, of course, broader than comic books and an important topic in itself.

Acknowledgement. All that I am, or ever hope to be, I owe to my angel mother.

References

1. Forchini, P.: Movie discourse: Marvel and DC studios compared. In: The Routledge Handbook of Approache Analysis, pp. 183–201. Routledge (2020)
2. Ricker, A.: Call it Science: Biblical Studies, Science Fiction, and the Marvel Cinematic Universe (2021)
3. Park, J.-H., Kim, H.-N.: A study on the success factors of Marvel game using Marvel IP. In: Proceedings of the Korean Society of Computer Information Conference, pp. 155–158. Korean Society of Computer Information (2021)
4. Shi, C., Yu, X., Ren, Z.: How the Avengers Assembled? Analysis of Marvel Hero Social Network. arXiv preprint arXiv:2109.12900 (2021)
5. Liu, L., Wang, Y., Xu, Y.: A practical guide to counterfactual estimators for causal inference with time-series cross-sectional data. arXiv preprint arXiv:2107.00856 (2021)
6. Wauchope, H.S., et al.: Evaluating impact using time-series data. Trends Ecol. Evol. **36**(3), 196–205 (2020)
7. Huang, Y., Fu, Z., Franzke, C.L.E.: Detecting causality from time series in a machine learning framework. Chaos Interdisc. J. Nonlinear Sci. **30**(6), 063116 (2020)
8. Weichwald, S., Jakobsen, M.E., Mogensen, P.B., Petersen, L., Thams, N., Varando, G.: Causal structure learning from time series: large regression coefficients may predict causal links better in practice than small p-values. In: NeurIPS Competition and Demonstration, pp. 27–36. PMLR (2020)
9. Cliff, O.M., Novelli, L., Fulcher, B.D., Shine, J.M., Lizier, J.T.: Exact inference of linear dependence between multiple autocorrelated time series. arXiv preprint arXiv:2003.03887 (2020)
10. Lim, B., Zohren, S.: Time-series forecasting with deep learning: a survey. Philos. Trans. R. Soc. A **379**(2194), 20200209 (2021)

A Decade of Artificial Intelligence Research in the European Union: A Bibliometric Analysis

Agata Frankowska[✉] and Bartosz Pawlik

National Information Processing Institute, Warsaw, Poland
{agata.frankowska,bartosz.pawlik}@opi.org.pl

Abstract. In recent years, the body of research on artificial intelligence (AI) has grown rapidly. As the European Union strives for excellence in AI development, this study aims to establish the publication achievements in the field among its member states between 2010 and 2019. We applied clustering and principal component analysis (PCA) on a set of bibliometric data concerning research publications on AI obtained from Scopus. The results reveal that while the union's most populous countries—the United Kingdom, Germany, France, Spain, and Italy—were the most prolific producers of AI publications between 2010 and 2019, the highest impact was noted for publications that originated in the Nordic and Benelux countries, as well as in Austria and Ireland. Analysis confirms that the division between 'old' and 'new' member states has endured: the nations that joined the EU after 2004 recorded the lowest results in scientific output and impact in the AI field. This study can assist research agencies and researchers in developing a broad grasp of the current state of AI research.

Keywords: Artificial intelligence · Bibliometric analysis · Bibliometric indicators · Clustering · European Union · Principal component analysis

1 Introduction

The body of research on artificial intelligence (AI) has grown rapidly in application and use. AI is soon likely to become a general purpose technology; this will affect the nature of work and the labour market [1, 2], and inequalities [3], as well as entailing significant economic [4] and societal impacts [5, 6]. Researchers in AI have long recognised the absence of a definitional consensus in the field [7, 8]. Policymakers and AI researchers have adopted different approaches in defining the term: while the former tend to favour human thinking or behaviour, the latter focus on technical problem specification and functionality [9]. A significant degree of conceptual development and controversy exists around the notion of AI; this has produced heated philosophical and ethical debates [10].

AI has garnered much attention in academic research, as well as in geopolitics. Although it is well known that the United States and China are the current leaders, the European Union (EU) harbours its own ambitions in the field [11]. Since 2017 [12], the EU has seen considerable development of the AI-relevant regulatory policies necessary to settle the scene for the technology's development. According to the EU White Paper

C. Biele et al. (Eds.): MIDI 2021, LNNS 440, pp. 52–62, 2022.
https://doi.org/10.1007/978-3-031-11432-8_5

on Artificial Intelligence—one component of the digital agenda package adopted by the European Commission in 2020—AI is broadly understood as '*a collection of technologies that combine data, algorithms and computing power*' [13]. The White Paper asserts that uptake of AI across sectors of the economy can support '*Europe's technological sovereignty*'. This, in turn, is conditioned by the development of the research and innovation sector. In order to boost research and innovation, the EU must undertake a wide range of interventions: increasing investment for AI, supporting basic and industrial research, creating a network of AI Research Excellence Centres, and delivering AI to companies through Digital Innovation Hubs – as well as establishing testing and experimentation infrastructures for AI. It is also crucial that AI be incorporated into higher education programmes and attract top-tier scholars. Moreover, the EU has obliged its member states to issue national AI development strategies, recognising that AI diffusion pat-terns will benefit from a conducive policy environment. As the EU is considered a global leader in regulation, policy solutions adopted inside the bloc can also influence the direction of AI regulation beyond its borders [12].

Against a backdrop of such plain ambition in the race for AI, this paper aims to address the question: how did European Union member states perform in AI research between 2010 and 2019? Answering this question is central to ascertaining the current state of AI scholarship and revealing insights into the research performed in the last decade. A response will also enable scientists and research agencies to lay plans for the future.

2 Method

Our approach applies unsupervised machine learning methods to classify EU member states into homogeneous groups based on bibliometric data[1]. The Scopus database— which, along with Web of Science, is one of the widely used repositories for performing bibliometric analyses [14]—served as the key data source. To retrieve AI-related publications from Scopus, we used English-language AI keywords. The initial list was drafted on the basis of sources listed in the *References* section. Preparation of the search terms was guided by the assumption that they should identify publications which concern the AI topic in a broad sense, i.e. take into account both technical and social aspects of the term. Next, the list was verified by six independent experts at the Polish National Information Processing Institute – each of whom has extensive hands-on experience of AI. At this stage, words which were not unique to the AI field were excluded. Experts had also the opportunity to present their own propositions of keywords. Then, the candidate keywords were used to search for publications. Of this dataset, a sample of titles and publication keywords was selected to verify the validity of the terms in identifying appropriate publications. After elimination of words which were too ambiguous (e.g. *systems theory, deep blue, blackboard system, CAD, finite differences, knowledge acquisition, neural networks, knowledge representation*), the final keyword list comprised 276 words related to AI theme.

---

[1] We treated the United Kingdom as part of the European Union, as it formally departed the bloc on January 31, 2020.

The publication data was retrieved from Scopus API on 15 April 2020. A publication was included in the dataset if it contained at least one of the 276 keywords in its title or keywords. After limiting our search to English-language articles and conference materials published between 2010 and 2019 by authors affiliated to institutions in EU member states, our dataset was restricted to 175 808 publications. It is noteworthy that each of the scientists could potentially be affiliated to multiple institutions.

Based on the information in the publication dataset, the following set of bibliometric indicators were calculated at national level:

(1) **Number of publications (p).** This metric measures a member state's total scientific output. If a publication has more than one author from different countries, we counted one for each country, i.e. full counting method was applied.

(2) **Number of citations per publication (cpp)**. The total number of citations divided by the total number of publications of a given member state. Due to the limitations of Scopus's API, self-citation publications were not excluded and an open-ended citation window was used. A publication was classified as cited if it scored at least one citation prior to the date the data was collected. As this indicator does not correct for differences in citations collected over time, differences in citation rates for different document types, nor field-specific differences in citation frequencies, the citation indicators described below (MNCS and pp_top) were normalised for publications of similar ages, types, and fields.

(3) **Mean normalised citation score (MNCS)** [15, 16]. The indicator normalises the citation impact of a given paper with the mean impact of similar papers published in the same field, document type, and publication year. In our analysis, the normalisation for field differences was based on Scopus's All Science Journal Classification (ASJC) system, which comprises 27 main fields. Each journal in which a publication appeared belongs to one or more subject areas. Since normalised indicators are fundamentally inconsistent when full counting is used [17, 18], our indicators were obtained using an author-level fractional counting method.

(4) **Proportion of member states' scientific output belonging to the 10% most-cited European Union documents (pp_top)** [19]. This indicator, which is also known as the excellence indicator, measures the proportion of a member state's research output that belongs to the top 10% most frequently cited, compared with others of the same field, publication year, and document type. Like MNCS, the normalisation for field differences is based on Scopus's ASJC. Contrary to mean-based indicators—which are constructed using extremely skewed citation distributions, and can be determined by a low number of frequently-cited documents—the excellence indicator is robust to outliers, as it exclusively considers publications at the high end (e.g. 10%) of the citation distribution. To overcome the difficulty of ties (i.e. publications having the same number of citations), we applied Waltman & Schreiber's method [20].

(5) **Proportion of member states' documents produced in international collaboration (pp_int)**. This metric captures the proportion of publications by co-authors who are affiliated to two or more countries.

On the basis of the above, cluster analysis was performed using a k-means algorithm to partition the EU member states into multiple and homogenous groups. Prior to the clustering, the variables were standardised and clustering tendency [21] was assessed using Hopkins statistic [22]. Its value was 0.74, which indicates the presence of significant and meaningful clusters. Due to the nondeterministic nature of k-means, the algorithm was initialised 50 times for different values of k. The value that minimised the objective function, within-cluster variation, was selected for further analysis. To determine the optimal number of clusters, we deployed the elbow method [23] with the support of the silhouette coefficient [24]. The latter was also used to evaluate the fit of individual observations in the classification and quality of clusters, as well as the broader classification.

3   Results

3.1   Clustering

The member states were grouped into three clusters, in accordance with both the elbow method and the silhouette coefficients. For evaluation of the clustering fit of individual observations, a silhouette plot is presented in Fig. 1. It illustrates the silhouette widths sorted in descending order computed for member states, which were grouped into three clusters obtained using the k-means algorithm. Each cluster is presented in a different colour. The grey dotted vertical line depicts the overall average silhouette width for whole classification.

The resulting partition contains the three main clusters:

- Cluster 1 comprises the five most populous countries in the EU: Germany, Italy, Spain, the United Kingdom, and France. Germany was classified with the greatest certainty, as confirmed by having the widest silhouette (0.55) of the five.
- Cluster 2 comprises three Nordic countries (Finland, Denmark, and Sweden), the Benelux countries (Belgium, Netherlands and Luxembourg), Ireland, and Austria. This cluster presents the most profound and visible structure (silhouette width = 0.60). Luxembourg was classified with the lowest certainty (silhouette width = 0.48); Finland was classified with the greatest – both within this cluster and among all EU member states (silhouette width = 0.69)
- Cluster 3 comprises Central and Eastern European countries (Hungary, Poland, Czechia, Croatia, Slovakia, Slovenia, Bulgaria, and Romania), the Baltic states (Estonia, Latvia, and Lithuania) and four Southern European countries (Malta, Greece, Portugal, and Cyprus). Poland, Slovakia, Greece, Portugal, Cyprus, and Estonia were classified with the lowest certainty, which is illustrated by their respective silhouette widths of less than 0.5. As Estonia recorded a silhouette width slightly above zero (0.04), it is considered an intermediate case, that lies between Clusters 2 and 3; determining which assignment is more valid is not unequivocal.

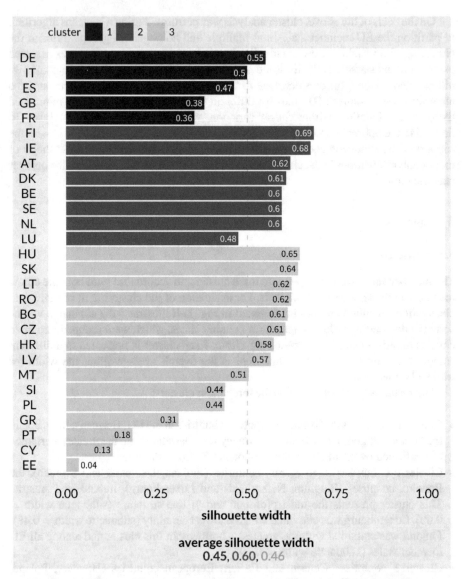

Fig. 1. Clusters of EU member states based on bibliometric indicators (1)–(5).

3.2 Principal Component Analysis

Having clustered the EU member states, we used principal component analysis (PCA) to reduce dimensionality among the highly correlated dataset and visualize it using a lower number of dimensions – principal components. Table 1 presents Spearman's rank correlations coefficients between bibliometric indicators used in clustering and dimensional reduction algorithms. According to Kaiser-Guttman rule, principal components with eigenvalues greater than one should be retained. As a result, two principal components out of five were retained, which accounted for more than 93% of original data's variability.

Table 1. The Spearman's rank correlations between the bibliometric indicators, $\alpha = 0.05$

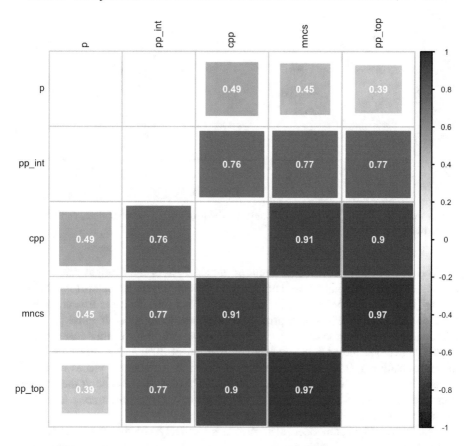

Figure 2 illustrates a biplot – a graphical representation of principal component scores and principal component loadings. The variables (bibliometric indicators) are plotted as black arrows, their loadings as coordinates, and the coloured circles as the principal component scores for the observations (EU member states). The distance between the variables' loadings and the plot origin represents the quality of each variable on the factor map. A supplementary categorical variable that represents each observation's assignment to a given cluster obtained from k-means was added to the plot and is depicted as a colour aesthetic.

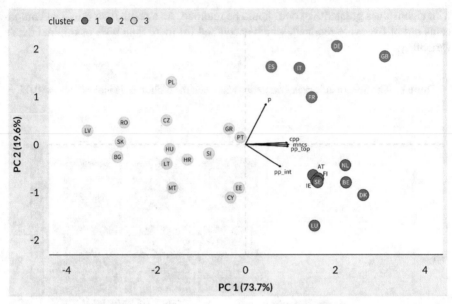

Fig. 2. A biplot of the bibliometric indicators (variables) and EU member states (observations) on the two principal components.

The first principal component captures almost 74% of the original data variance and separates member states based primarily on citation impact indicators (cpp, MNCS, and pp_top). They correlate highly and positively not only with the first component, but also with each other, as their principal component loadings lie in close proximity. Although the contribution made to the first component spreads evenly across citation impact indicators, the MNCS—with a contribution of 30% and a very high quality of representation at 0.98—is the variable on which the component places most of its weight. This means that it is the best-suited variable for approximating the scientific impact dimension, and can be considered a 'consensus' measure among all citation impact indicators used in the PCA.

The first principal component also correlates positively with the member states' proportions of international publications (pp_int). According to the Spearman's rank correlation coefficients reported in Table 1, a moderate, positive relationship exists between citation impact indicators and member states' degrees of international collaboration; thus, publications of authors affiliated to countries whose scientific output relies heavily on foreign cooperation are characterised by greater impacts.

The member states of Clusters 1 and 2, which are located on the right side of the Y axis, perform well on citation impact indicators; the opposite is true for those positioned on the left (Cluster 3). The fourth quadrant contains member states that scored relatively highly on scientific impact indicators, and had high proportions of international publications in their AI research output. These countries can be found in Cluster 2.

The second principal component retains more than 19% of dataset's variability and puts weight on the total number of publications (p); approximately 80% of the second component is influenced solely by this metric. The remainder is attributed to the scientific collaboration indicator; this is reflective of the degree of foreign cooperation, with which it correlates negatively. International collaboration has no significant effect on the size of member states' publication output. This is reflected by the perpendicular placement of their loadings, as illustrated in the biplot. For these reasons, our results differ from the findings of Aksnes et al. [17], who observed a negative correlation between member states' scientific output and the degree to which they collaborate internationally.

The member states with the largest scientific outputs in AI—are placed above the X axis, leaning towards the first quadrant; conversely, the lowest ranked member states are positioned in the third quadrant. It is noteworthy that while all countries in Cluster 3 generally achieved low scores in impact indicators, those located in the second quadrant had relatively large AI research outputs. One member state within this cluster, Portugal, achieved average scores on all indicators, as it is located near the centre of the biplot.

4 Summary and Conclusions

The primary aim of this study was to construct an overview of the AI publication achievements in the European Union member states between 2010 and 2019. We have presented a clear division of member states into three clusters on the basis of a set of bibliometric indicators that reflect scientific output, impact, and collaboration. Clusters 1 and 2 comprised solely 'old' member states, while Cluster 3 mostly comprised 'new' ones that joined the EU after 2004 – with the exceptions of Portugal and Greece. The member states of Cluster 1—the United Kingdom, Germany, France, Italy, and Spain—performed strongly in AI research output between 2010 and 2019. These results, however, correlate with the sizes of those member states' populations. Concerning scientific impact, the member states contained in Cluster 2 achieved the best results – even without having the largest scientific outputs. It should be noted that the citation impact of these countries might also relate to internationalisation. Twelve Central, Eastern, and Southern member states, together with the three Baltic states comprised Cluster 3 – the largest one. Those member states' achievements in scientific output, impact, and collaboration are far lesser than those of Clusters 1 and 2.

The analysis allows us to draw several conclusions. First, our results confirm that AI research is a field in which the division between 'old' and 'new' EU member states has endured since the 2004 accession. With evidence that the United Kingdom was the top producer of AI publications between 2010 and 2019, this paper supports the conclusion that the United Kingdom's exit from the bloc will influence the EU's future scientific performance in AI considerably [25].

Second, our analysis focuses on EU member states, but their position in the AI research should be viewed in a broader geopolitical context: one that incorporates the AI aspirations of the United States and China. According to Scopus data, between 2010 and 2019, the largest number of AI publications was produced by China, followed by the EU and the United States [26]. While studies have demonstrated clearly that the United States and the EU outperform China in terms of scientific impact (measured by the Field-Weighted Citation Impact, FWCI) [27] and of excellence (measured by the share of AI-related documents in the 10% most cited publications worldwide) [28], China has registered a remarkable increase in both measures in the field of AI; the country's FWCI rose from 0.6 in 2012 to 0.9 in 2017 – almost reaching the global average [11]. Moreover, China surpassed the EU for the first time in 2016, and the United States in 2020 [29] in its share of global AI journal citations and experienced an increase in scientific excellence between 2006 and 2016 [28]. This signals that China's position in AI research is likely to continue elevating.

Finally, sources indicate that beyond the three leaders—the United States, China, and the EU—other nations (most notably India) have also made progress in AI [28]. This serves to complicate the picture in which the United States, China, and the EU fully dominate the AI research landscape. While it can be envisaged that the domination of the three leaders will endure, and that there is a basis to compare such large entities due to their scales, questions remain on the extent of their future success, and whether newcomers in AI research can achieve comparably.

References

1. Acemoglu, D., Restrepo, P.: Automation and new tasks: how technology displaces and reinstates labor. J. Econ. Perspect. **33**, 3–30 (2019). https://doi.org/10.1257/jep.33.2.3
2. Lane, M., Saint-Martin, A.: The impact of Artificial Intelligence on the labour market (2021). https://www.oecd-ilibrary.org/content/paper/7c895724-en
3. Korinek, A., Stiglitz, J.E.: Artificial Intelligence and Its Implications for Income Distribution and Unemployment. National Bureau of Economic Research, Inc. (2017)
4. Agrawal, A., Gans, J., Goldfarb, A.: Economic Policy for Artificial Intelligence (2018)
5. Omar, M., Mehmood, A., Choi, G.S., Park, H.W.: Global mapping of artificial intelligence in Google and Google Scholar. Scientometrics **113**(3), 1269–1305 (2017). https://doi.org/10.1007/s11192-017-2534-4
6. Stone, P., Brooks, R., Brynjolfsson, E., Calo, R., Etzioni, O., Hager, G., et al.: Artificial intelligence and life in 2030. Stanford University (2016)
7. Kaplan, A., Haenlein, M.: Siri, Siri, in my hand: Who's the fairest in the land? On the interpretations, illustrations, and implications of artificial intelligence. Bus. Horiz. **62**, 15–25 (2019). https://doi.org/10.1016/j.bushor.2018.08.004
8. Russel, S.J., Norvig, P.: Artificial Intelligence. A Modern Approach. Pearson Education (2016)
9. Krafft, P.M., Young, M., Katell, M., Huang, K., Bugingo, G.: Defining AI in Policy versus Practice. Proc. 2020 AAAIACM Conf. AI Ethics Soc. AIES. (2020)
10. Stahl, B.C., Antoniou, J., Ryan, M., Macnish, K., Jiya, T.: Organisational responses to the ethical issues of artificial intelligence. AI Soc. 1–15 (2021). https://doi.org/10.1007/s00146-021-01148-6
11. Castro, D., McLaughlin, M., Chivot, E.: Who Is Winning the AI Race: China, the EU or the United States? Center for Data Innovation (2019)
12. Andraško, J., Mesarčík, M., Hamuľák, O.: The regulatory intersections between artificial intelligence, data protection and cyber security: challenges and opportunities for the EU legal framework. AI Soc. **36**(2), 623–636 (2021). https://doi.org/10.1007/s00146-020-01125-5
13. European Commission: White Paper on Artificial Intelligence - A European approach to excellence and trust, (2020)
14. Zhu, J., Liu, W.: A tale of two databases: the use of Web of Science and Scopus in academic papers. Scientometrics **123**(1), 321–335 (2020). https://doi.org/10.1007/s11192-020-03387-8
15. Waltman, L., van Eck, N.J., van Leeuwen, T.N., Visser, M.S., van Raan, A.F.J.: Towards a new crown indicator: an empirical analysis. Scientometrics **87**, 467–481 (2011). https://doi.org/10.1007/s11192-011-0354-5

16. Rehn, C., Wadskog, D., Gornitzki, C., Larsson, A.: Bibliometric Indicators - Definitions and usage at Karolinska Instututet. Karolinska Institutet (2014)
17. Aksnes, D., Schneider, J., Gunnarsson, M.: Ranking national research systems by citation indicators. A comparative analysis using whole and fractionalised counting methods. J Informetr. **6**, 36–43 (2012). https://doi.org/10.1016/j.joi.2011.08.002
18. Waltman, L., van Eck, N.J.: Field-normalized citation impact indicators and the choice of an appropriate counting method. J. Informetr. **9**, 872–894 (2015). https://doi.org/10.1016/j.joi.2015.08.001
19. Leydesdorff, L., Bornmann, L., Mutz, R., Opthof, T.: Turning the tables on citation analysis one more time: Principles for comparing sets of documents. J. Am. Soc. Inf. Sci. Technol. **62**, 1370–1381 (2011). https://doi.org/10.1002/asi.21534
20. Waltman, L., Schreiber, M.: On the calculation of percentile-based bibliometric indicators. J. Am. Soc. Inf. Sci. Technol. **64**, 372–379 (2013). https://doi.org/10.1002/asi.22775
21. Jain, A.: Data clustering: 50 years beyond K-means. Award Win. Pap. 19th Int. Conf. Pattern Recognit. ICPR. **31**, 651–666 (2010). https://doi.org/10.1016/j.patrec.2009.09.011
22. Lawson, R.G., Jurs, P.C.: New index for clustering tendency and its application to chemical problems. J. Chem. Inf. Comput. Sci. **30**, 36–41 (1990). https://doi.org/10.1021/ci00065a010
23. Syakur, M.A., Khotimah, B.K., Rochman, E.M.S., Satoto, B.D.: Integration K-means clustering method and elbow method for identification of the best customer profile cluster. IOP Conf. Ser. Mater. Sci. Eng. **336**, 012017 (2018). https://doi.org/10.1088/1757-899x/336/1/012017
24. Rousseeuw, P.J.: Silhouettes: a graphical aid to the interpretation and validation of cluster analysis. J. Comput. Appl. Math. **20**, 53–65 (1987). https://doi.org/10.1016/0377-0427(87)90125-7
25. European Commission: Science, Research and Innovation Performance of the EU 2020. A fair, green and digital Europe. Publications Office of the European Union, Luxembourg (2020)
26. Frankowska, A., Pawlik, B., Feldy, M., Witkowska, E.: Artificial Intelligence and science and higher education sector. Review of strategic documents and scienfitic achievements [PL: Sztuczna inteligencja a sektor nauki i szkolnictwa wyższego. Przegląd dokumentów strategicznych i osiągnięć na świecie]. National Information Processing Institute, Warszawa (2020)
27. Elsevier: Artificial Intelligence: How knowledge is created, transferred, and used. Trends in China, Europe and the United States (2018)
28. Correia, A., Reyes, I.: AI research and innovation: Europe paving its own way. Europaen Commission. Directorate-General for Research and Innovation (2020)
29. Zhang, D., et al.: The AI Index 2021 Annual Report. Human-Centered AI Institute, Stanford University, Stanford, CA, AI Index Steering Committee (2021)

Sources used in the keyword selection:

30. Association for the Advancement of Artificial Intelligence, AITopics. https://aitopics.org/search
31. China Institute for Science and Technology Policy at Tsinghua University: China AI Development Report (2018). http://www.sppm.tsinghua.edu.cn/eWebEditor/UploadFile/Executive_susmmary_China_AI_Report_2018.pdf
32. Corea F.: AI Knowledge Map: How To Classify AI Technologies. https://www.forbes.com/sites/cognitiveworld/2018/08/22/ai-knowledge-map-how-to-classify-ai-technologies/#35a4feaf7773
33. Glossary of Artificial Intelligence. https://en.wikipedia.org/wiki/Glossary_of_artificial_intelligence

34. Goodfellow, I., Bengio, Y., Courville, A.: Deep Learning (2018). https://github.com/janishar/mit-deep-learning-book-pdf/blob/master/complete-book-pdf/deeplearningbook.pdf

Localization Using DeepLab in Document Images Taken by Smartphones

Shima Baniadamdizaj$^{(\boxtimes)}$ ⓘ

Department of Electrical and Computer Engineering, Kharazmi University, Tehran, Iran
baniadam.shima@gmail.com

Abstract. The seamless integration of statistics from virtual and paper files could be very crucial for the know-how control of efficient. A handy manner to obtain that is to digitize a report from a picture. This calls for the localization of the report in the picture. Several approaches are deliberate to resolve this hassle; however, they are supported historical picture method strategies that are not robust to intense viewpoints and backgrounds. Deep Convolutional Neural Networks (CNNs), on the opposite hand, have been validated to be extraordinarily strong to versions in heritage and perspective of view for item detection and classification duties. Inspired by their robustness and generality, we advocate a CNN-primarily based totally technique for the correct localization of files in real-time. We advocate the new utilization of Neural Networks (NNs) for the localization hassle as a key factor detection hassle. The proposed technique ought to even localize snapshots that don't have a very square shape. Also, we used a newly amassed dataset that has extra tough duties internal and is in the direction of a slipshod user. The result is knowledgeable in 3 specific classes of snapshots and our proposed technique has 100% accuracy on easy one and 77% on average. The result is as compared with the maximum famous report localization strategies and cell applications.

Keywords: Document corner localization · Smartphone image capturing · Image dataset · Image processing

1 Introduction

Based on the convenience of use and portability of smartphones, enhancing the processing power, and enhancing the first-class of pictures taken the usage of smartphones, those telephones had been capable of partly updating the paintings of scanners in file imaging. In the meantime, because of the distinctive abilities of smartphones and scanners, there are issues and demanding situations alongside the manner of turning telephones into scanners. Also, scanners are slow, costly, and now no longer portable. Smartphones, on the other hand, have become very accessible.

There are distinctive kinds of file paper files that are simpler to carry, read, and proportion while virtual files are simpler to search, index, and save. One of the benefits of file imaging is the cap potential to transform a paper file right into a virtual file, which through persevering with the technique of digitization and the usage of digital

C. Biele et al. (Eds.): MIDI 2021, LNNS 440, pp. 63–74, 2022.
https://doi.org/10.1007/978-3-031-11432-8_6

letter reputation strategies, the textual content, and contents of the file may be without difficulty edited and searched. It is likewise viable to save the virtual file withinside the area of outside reminiscence and without difficulty proportion or switch it among humans with a smaller extent than the distance that a paper file occupies withinside the environment.

There are a few demanding situations to digitizing a file the usage of a cellphone, some of which can be stated. Lack of uniform light, shadows at the file, which may be the shadow of the hand, phone, or different objects, form of materials, colors, and functions of the file, version withinside the historical past of the file and its contents, having three-D distortion, blurring, historical past complexity of files (which includes covered pages, chessboard, etc.), low file contrast, or terrible cellphone digital digicam first-class, undetectable life of the file from its historical past (because of being the equal color, light, etc.), the complexity of the file, for example, having folds, taking pix of multi-web page files which includes books and identification cards, being a part of the file out of the picture, covered. Being a part of the file through different objects, etc.). The best approach ought to be strong in those demanding situations. It needs to also be capable of running on a cellphone in a completely less expensive time.

In general, withinside the subject of digitizing files, a few researchers had been withinside the subject of resolving or supporting to enhance the picture first-class and decreasing the issues stated withinside the preceding paragraph, and others have supplied algorithms that during the case of issues withinside the picture taken through the careless person, the file can nonetheless be located withinside the picture. There is a 3rd class of studies that, whilst enhancing picture first-class and guiding the person to seize the best first-class picture of the file, offers the set of rules had to discover the file withinside the picture, that is an aggregate of the preceding strategies.

We advise a way that makes use of deep convolutional neural networks for semantic segmentation in pictures taken through smartphones. Our approach outperforms the kingdom of the artwork in the literature on three-D distorted pictures and might run in real-time on a cellphone. Additionally, it's miles extraordinary than previous strategies with inside the feel that its miles custom-designed to be robust to extra problems certainly through education on extra consultant data.

2 Literature Review

2.1 Document Localization Datasets

To localize files in photographs taken via way of means of smartphones we want a real-global dataset this is amassed from an ordinary user. There are four special datasets in report photographs taken via way of means of smartphones task. Three of those datasets comprise photographs that have identical or very near photographs together. The fourth dataset amassed extra photographs than others and changed into additionally in the direction of the real-global taken photographs with numerous demanding situations.

The to be had information set changed into used for the qualitative assessment of photographs of files excited about smartphones [1]. The information set of Kumar et al. Accommodates 29 special files below special angles and with blurring, and finally, 375 photographs had been obtained. The dataset offered in [2] makes use of 3 not unusual place forms of paper in doing away with numerous forms of distortion or harm consisting of blurring, shaking, special lights conditions, combining forms of distortion in an image, and taking photographs that have one or extra distortions on the identical time, and the usage of numerous forms of smartphones, which makes this information set extra reliable. The information set is offered withinside the article [3], which covers a few factors of the scene, consisting of the light's conditions. An easy historical past changed into used. A robot arm was changed into used to take photographs to dispose of the digital digicam shake. The identical concept [4] offered a video dataset that there are five classes from easy to complicated, all with the identical content material and historical past, it consists of movies with 20 frames. And photographs are extracted from those frames. Different smartphones had been used for the harm resulting from the device, and additionally via way of means of the usage of special files. An overall of four, 260 special photographs of 30 files had been taken.

Paper [5] gives a Mobile Identification Document Video information set (MIDV-500) inclusive of 500 movies for fifty precise identity report sorts with floor truth, permitting studies on an extensive type of report processing issues. The paper gives capabilities of the information set and assessment consequences for present techniques of face recognition, textual content line recognition, and information extraction from report fields. Because the sensitivity of identity papers, which consist of non-public information, is a critical aspect, all photographs of supply files utilized in MIDV-500 are both withinside the public area or launched below public copyright licenses. In the paper [6] a brand-new report dataset is offered this is in the direction of the real-global photographs taken via way of means of users. The information is labeled into easy, middle, and complicated responsibilities for detection. It consists of nearly all demanding situations and consists of numerous report sizes and brands and backgrounds. It compares the result of the report localizing techniques with famous techniques and cell applications.

2.2 Document Localization Methods

Due to the demanding situations, it isn't always viable to digitize files with the use of smartphones without preprocessing or post-processing and count on suitable outcomes in all situations. That is why algorithms had been proposed to enhance the outcomes. The impact of image venture algorithms on the result may be divided into 3 categories: 1. Reduce demanding situations earlier than capturing 2. Fixed problems even as taking snapshots 3. Solve demanding situations after capturing. One of the earliest strategies of record localization changed into primarily based totally on a version of the history for segmentation. The history changed into modeled with the aid of using taking a photo of the history without the record. The distinction between the 2 photographs changed into applied to decide in which the record changed into found. This approach had the apparent negative aspects that the digital digicam needed to be stored desk-bound and pictures needed to be taken [7]. In general, the algorithms used to locate the record withinside the photo may be divided into 3 categories: 1. Use of extra hardware 2. Depend upon photo processing strategies 3. Use of system studying techniques. This trouble has arisen with the unfold of smartphones from 2002 to 2021 and may be improved.

2.2.1 Additional Hardware

In the article [8] they gift courses for the consumer in taking with fewer demanding situations primarily based totally on exceptional functions. As a result, the photograph calls for tons much less pre-processing to localize the document. This technique became now no longer very consumer-pleasant for the customers because of the constraints and slowdown of digitization. Article [9] used this technique for localizing. Following pre-processing, in addition, algorithms are required to complete the localization task. These algorithms can be divided into categories: 1. Use of more hardware 2. Consider strategies to apply gadgets imaginative and prescient. 3. The utility of deep gaining knowledge of algorithms. A scanning utility is presented [10] that consists of real-time web page recognition, fine assessment, and automated detection of a web page cover [11] while scanning books. Additionally, a transportable tool for putting the smartphone all through scanning is presented. Another paper that used extra hardware introduces a scale-invariant characteristic remodel into the paper detection machine [12]. The hardware of the paper detection machine includes a virtual sign processor and a complicated programmable good judgment tool. The equipment can obtain and process images. The software program of this machine makes use of the SIFT technique to discover the papers. Compared to the conventional technique, this set of rules offers higher with the detection process. In the paper [13] paper detection desires a sheet of paper with a few styles revealed on it. It takes a laptop imaginative and prescient era one step toward getting used withinside the field.

2.2.2 Machine Vision Techniques

The set of rules [14] operates with the aid of using finding ability line segments from horizontal test strains. Detected line segments are prolonged or merged with neighboring test line textual content line segments to provide larger textual content blocks, which might be finally filtered and refined. The paper [15] provides a spotting device in complicated history video frames. The proposed morphological method [16] is insensitive to noise, skew, and textual content orientation. So it's miles without artifacts as a result of each constant/most excellent international thresholding and constant-length block-primarily based nearby thresholding. [17] proposes a morphology-primarily based totally method for extracting key assessment traits as guidelines for attempting to find appropriate license plates. Preferred license plates. The assessment function is in lighting fixtures adjustments and invariant to numerous variations like scaling, translation, and skewing. The paper [18] applies aspect detection and makes use of a low threshold to filter non-textual content edges. Then, a nearby threshold is chosen to each hold low-assessment textual content and simplify the complicated history of the excessive assessment textual content content. Following that, textual content-region enhancement operators are proposed to emphasize areas with excessive aspect power or density.

[19] describes a step-with the aid of using-step method for locating candidate areas from the entered photo the use of gradient data, figuring out the plate region of the various candidates, and enhancing the region's border with the aid of using including a plate template. In the textual content extracts from video frames paper [20] the nook factors of the chosen video frames are detected. After deleting a few remoted corners, it

merges the closing corners to shape candidate textual content areas. Target frames [21] are decided on at constant time durations from photographs detected with the aid of using a scene-extrude detection approach. A color histogram is used to perform segmentation with the aid of using color clustering surrounding color peaks for every selection.

The approach [22] locates candidate areas without delay withinside the DCT compressed area by the use of the depth version data encoded withinside the DCT area. The paper [23] makes use of a clean history for pix to locate the areas of interest (ROI). [24] proposes a linear-time line phase detector with dependable findings, a constrained quantity of fake detections, and no parameter tweaking. On a big set of herbal pics, our approach is evaluated and as compared in opposition to present-day techniques. Using Geodesic Object Proposals [26], a technique for detecting ability files [25] is defined in a given photo. The entered pics had been downsampled to the useful resource with the extraction of structures/functions of interest, to lessen noise, and to enhance runtime pace and accuracy. The outcomes indicated that the use of Geodesic Object Proposals withinside the file item identity activity is promising. Also, [27] operators are associated with the max-tree and min-tree representations of files in pix. In paper [28] a simple-to-write set of rules is proposed to compute the tree of shapes; When statistics quantization is low, it works for nD pics and has a quasi-linear complexity.

The methodology [29] is primarily based totally on projection profiles blended with a linked aspect labeling process. Signal cross-correlation is likewise used to affirm the detected noisy textual content regions. Several awesome steps are used for this challenge [30] a pre-processing system the use of a low-by skip Wiener filter, a difficult estimation of foreground areas, a history floor computation with the aid of using interpolating adjoining history intensities, the edge cost with the aid of using combining the computed history floor with the authentic photo consisting of the pinnacle test of the photo, and sooner or later a post-processing step to enhance the nice of the textual content regions and keep line connectivity. Removing the skew impact on digitalized files [31] proposed that every horizontal textual content line intersects a predefined set of vertical strains at non-horizontal positions. Just with the aid of using the use of the pixels on such vertical strains, we create a correlation matrix and calculate the file's skew attitude with terrific precision. In the challenge of white beard files [32] evolved a strong function-primarily based totally method to routinely sew a couple of overlapping pix. The cautioned approach [33] is primarily based totally on the combinatorial production of ability quadrangle picks from hard and fast line segments, in addition to projective file reconstruction with a recognized focal length. For line detection, the Fast Hough Transform [34] is applied. With the set of rules, a 1D model of the brink detector is presented. Three localization algorithms are given in an article [35]. All algorithms employ function factors, and of them moreover have a take a observe near-horizontal and near-vertical strains at the photo. The cautioned method [36] is a distinctly specific file localization approach for spotting the file's 4 nook factors in herbal settings. The 4 corners are about expected withinside the first step the use of a deep neural network-primarily based Joint Corner Detector (JCD) with an interesting mechanism, which makes use of a selective interest map to the kind of discover the file location.

2.2.3 Machine Learning Method

The paper [37] suggests a CNN primarily based approach as it should be localized files in real-time and version localization hassle as a key factor detection hassle. The 4 corners of the files are collectively expected with the aid of using a Deep Convolutional Neural Network. In the paper [38] first, stumble on the sort of the file and classify the images, after which with the aid of using understanding the sort of the file a matched localization approach is finished at the file and allows data extraction. Furthermore, another method offered a new usage of U-Net for file localization in pictures taken through smartphones [39].

3 Methodology

We version the hassle of record localization as key factor detection. The approach desires a ground truth as a mask of the record component and the non-record component. We exhibit the record with white (255) and non-record components with black (0).

3.1 Dataset Preparation

We use [1–5] datasets as train and validation datasets and make the photo length comparable to the usage of the max photo top and width amongst snapshots the usage of 0 paddings. And use [6] dataset because of the check dataset for the assessment and evaluating the proposed approach with the preceding strategies and cellular applications.

3.2 Using Deep Neural Networks

For the mission of report localization in pix taken via way of means of smartphones, we used deeplabv3 [40] technique and the fine-tuning technique to retrain a few ultimate layers in the DeepLab neural network. This community has benefited from deconvolution. In this dissertation, we've taken into consideration locating the location of the report withinside the pix as a semantic segmentation mission. For convolutional neural networks, they've proven incredible overall performance in semantic photograph segmentation.

We use the possibilities of deeplabv3 to lessen the operational complexity via way of means of the usage of numerous neural community fashions withinside the semantic segmentation component like MobileNet [41]. The deeplabv3 with the mobilenetv2 has 2.11M parameters in total. We use mobilenetv2 as a function extractor in a simplified version of deeplabv3 to allow on-tool semantic segmentation. The resultant version achieves equal overall performance to the usage of mobilenetv1 as a function extractor (Fig. 1).

There are three layers in the DeepLab network. First, we emphasize atrous convolution [42], or convolution with up-sampled filters, as a useful technique in dense prediction problems. Throughout Deep Convolutional Neural Networks, atrous convolution allows us to directly regulate the resolution at which feature responses are calculated. It also enables us to efficiently expand the field of view of filters to include more context without increasing the number of parameters or calculation time.

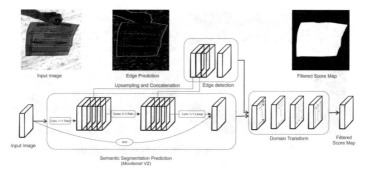

Fig. 1. Deeplabv3 using mobilenetv2 neural network architecture

Second, we present an atrous spatial pyramid pooling (ASPP) method for segmenting objects at different sizes with high accuracy. ASPP uses filters at different sample rates and functional fields-of-views to probe an input convolutional feature layer, collecting objects and picture context at different levels. Third, we combine approaches from DCNNs and probabilistic graphical models to enhance object boundary localization. In CNN's, the frequently used combination of max-pooling and downsampling produces invariance but at the cost of accuracy in localization, we solve this by integrating the answers at the final DCNN layer with a fully linked Conditional Random Field (CRF), which has been proven to increase localization accuracy both qualitatively and statistically.

Because of the problem of semantic segmentation, there are large and small instances that need to be segmented. If convolution kernels of equal size are used, the problem may arise that the receptive field is not large enough and the accuracy of segmentation of large objects may decrease. As a reaction to this problem, at the historic moment an atrous convolution was created, i.e. The size of the dilation rate was adjusted to modify the convolution kernel's receptive field. The impact of atrous convolution on a branch convolutional neural network, on the other hand, is not beneficial. If we continue to use smaller atrous convolutions to recover the information of small objects, a large redundancy will result. ASPP uses the dilation rate of different sizes to capture information on different scales in the network decoder. Each scale is an independent branch. It is merged at the end of the network and then a convolution layer is an output to predict the label. This approach successfully eliminates the gathering of unnecessary data on the encoder, allowing the encoder to focus just on the object correlation.

The training level needs a different ground truth from the paper [6] so we provide a masked ground truth (Fig. 2) that the document part and the non-document part Are differentiated with black and white colors. The document with white (255) and the non-document part with black (0). After freezing the intended layers, the final network has been updated and implemented in Ubuntu Linux version 16.04 LTS implementation and programming. The STRIX-GTX1080-O8G graphics card and the Core i7 6900k processor with 32 GB of RAM are also used for training, power, and network testing.

Fig. 2. Image sample with masked ground truth

4   Experiments and Results

4.1   Evaluation Protocol

The IoU method described in [43], has been used to evaluate. First, the perspective effect is deleted from the ground truth (G) and predicted (S) with the help of image size. We call new situations (G′) and (S′) respectively so that the IoU or Jaccard index is equal to:

$$IoU = (area(G' \cap S'))/(area(G' \cup S')) \tag{1}$$

The final result is the average of the IoU value for each image.

4.2   Results

The result is compared with all well-known methods, algorithms, and mobile applications that can solve the document localization task in images. The other methods' results are compared based on [6] (Fig. 3).

Fig. 3. The result of the proposed method on dataset [6]

In Table 1, the final results in different categories are presented and Fig. 4 shows the result in comparison with the previous methods. We run the model on a check the dataset and look at our outcomes to the previously released outcomes on the same dataset [6]. While this check is sufficient to evaluate the tool's accuracy on the competition dataset, it does have drawbacks. One, it cannot show to us how effective our tool conflates to unknown contents because of the reality several samples of contents were used for education. Second, its miles now no longer capable of providing records about how well our framework generalizes to unseen documents with similar content material fabric.

We cross-validate our technique through manner of way of deleting each content material fabric from the education setting and then checking on the deleted content material fabric to recovery the weaknesses and diploma generalization on unseen content material fabric. The quit result is in comparison with all well-known methods, algorithms, and mobile packages that might solve the document localization venture in snapshots. The special methods' outcomes regions compared, based definitely on [6] (Fig. 3). Our technique successfully generalizes to unseen documents, as outcomes are validated in Fig. 4.

This is not unreasonable for the purpose that the low choice of the entered picture graph prevents our model from relying on features of the document's layout or content material fabric. We moreover end from the outcomes that our technique generalizes well to unseen smooth contents. It is crucial to mention that the technique is designed to be effective on a low huge style of reasserts like a midrange smartphone without the usage of cloud or server-based recourses. The frames from the four corners are processed within the order within the implementation. It can implement snapshots all at once through a manner of way of taking walks a batch of four snapshots via the model. This must result in a substantial boom in inefficiency.

Table 1. The result of the proposed method on dataset [6]

Simple	Medium	Difficult	Average
100%	74%	58%	77%

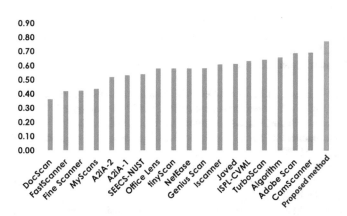

Fig. 4. The result of the proposed method compared with previous methods

5 Conclusion

In this paper, we provided a new application of DeepLabv3 the use of MobileNetv2 for report localization in pics taken with the aid of using smartphones. The very last result is the exceptional result of those tasks' methods. We used all dependable datasets on this task. And the primary dataset to evaluate is the newly gathered dataset with a diverse variety of report localization challenges. And finally, we offer a utility the use of the Kivy framework.

We moreover go through some sensible techniques that can be carried out with the usage of software program software belongings like Python, PyTorch, TensorFlow, and OpenCV. We used all reliable datasets in this task. And, the number one dataset to study is the newly accumulated dataset with a various variety of document localization challenges. Also, we gift a unique method for locating documents in pictures. The problem of localization is modeled as a problem of key element detecting. We show that this method should make assumptions nicely on new and unseen documents via the usage of a deep convolutional network.

References

1. Sheikh, H.R., Sabir, M.F., Bovik, A.C.: A statistical evaluation of recent full reference image quality assessment algorithms. IEEE Trans. Image Process. **15**(11), 3440–3451 (2006)
2. Ye, P., Doermann, D.: Document image quality assessment: a brief survey. In: 2013 12th International Conference on Document Analysis and Recognition (ICDAR). IEEE (2013)
3. Nayef, N., et al.: SmartDoc-QA: a dataset for quality assessment of smartphone captured document images-single and multiple distortions. In: 13th International Conference on Document Analysis and Recognition (ICDAR). IEEE (2015)
4. Burie, J.C., Chazalon, J., et al.: ICDAR2015 competition on smartphone document capture and OCR (SmartDoc). In: 13th International Conference on Document Analysis and Recognition, IEEE (2015)
5. Arlazarov, V.V., et al.: MIDV-500: a dataset for identity document analysis and recognition on mobile devices in video stream. Компьютерная оптика **43**.5 (2019)
6. Dizaj, S.B., Soheili, M., Mansouri, A.: A new image dataset for document corner localization. In: 2020 International Conference on Machine Vision and Image Processing (MVIP). IEEE (2020)
7. Lampert, H., Braun, H.T., et al.: Oblivious document capture and real-time retrieval. In: Proceedings. Camera-Based Document Analysis and Recognition, pp. 79–86 (2005)
8. Chen, F., et al.: SmartDCap: semi-automatic capture of higher quality document images from a smartphone. In: Proceedings of the 2013 International Conference on Intelligent User Interfaces. ACM (2013)
9. Jayaraman, D., et al.: Objective quality assessment of multiply distorted images. In: 2012 Conference Record of the Forty Sixth Asilomar Conference on Signals, Systems and Computers (ASILOMAR). IEEE (2012)
10. Kleber, F., et al.: Mass digitization of archival documents using mobile phones. In: Proceedings of the 4th International Workshop on Historical Document Imaging and Processing. ACM (2017)
11. Fototechnischer Ausschuss der KLA. 2016. Wirtschaftliche Digitalisierung in Archiven. (2016)
12. Zhu, J., Wang, S., Meng, F.: SIFT method for paper detection system. In: 2011 International Conference on Multimedia Technology (ICMT), IEEE (2011)
13. Quan, N., Zhou, X., Chen, X.: Scan paperback books by a camera. In: 2016 IEEE International Conference on Information and Automation (ICIA) (2006)

14. Zunino, R., Rovetta, S.: Vector quantization for license-plate location and image coding. IEEE Trans. Industr. Electron. **47**(1), 159–167 (2000)
15. Kuwano, H., et al.: Telop-on-demand: video structuring and retrieval based on text recognition. In: 2000 IEEE International Conference on Multimedia and Expo. ICME2000. Proceedings. Latest Advances in the Fast-Changing World of Multimedia (Cat. No. 00TH8532). vol. 2. IEEE (2002)
16. Hasan, Y.M.Y., Karam, L.J.: Morphological text extraction from images. IEEE Trans. Image Process. **9**(11), 1978–1983 (2000)
17. Hsieh, J.-W., Yu, S.-H., Chen, Y.-S.: Morphology-based license plate detection from complex scenes. In: Object Recognition Is Supported by User Interaction for Service Robots. vol. 3. IEEE (2002)
18. Cai, M., Song, J., Michael, R., Lyu, A.: A new approach for video text detection. In: Proceedings of the International Conference on Image Processing. vol. 1. IEEE (2002).
19. Kim, S., et al.: A robust license-plate extraction method under complex image conditions. In: Object Recognition Is supported by User Interaction for Service Robots. vol. 3. IEEE (2002)
20. Hua, X.-S., et al.: Automatic location of text in video frames. In: Proceedings of the 2001 ACM Workshops on Multimedia: Multimedia Information Retrieval. ACM (2001)
21. Kim, H.-K.: Efficient automatic text location method and content-based indexing and structuring of video database. J. Vis. Commun. Image Represent. **7**(4), 336–344 (1996)
22. Zhong, Y., Zhang, H., Jain, A.K.: Automatic caption localization in compressed video. IEEE Trans. Pattern Anal. Mach. Intell. **22**(4), 385–392 (2000)
23. Wu, V., Manmatha, R., Riseman, E.M.: Finding text in images. In: ACM DL (1997)
24. Gioi, V., Grompone, R., et al.: LSD: A fast line segment detector with a false detection control. IEEE Trans. Pattern Anal. Mach. Intell. **32**(4), 722–732 (2010)
25. Leal, L.R.S., Bezerra, B.L.D.: Smartphone camera document detection via geodesic object proposals. In: 2016 IEEE Latin American Conference on Computational Intelligence (LA-CCI), IEEE (2016)
26. Krähenbühl, P., Koltun, V.: Geodesic object proposals. In: Fleet, D., Pajdla, T., Schiele, B., Tuytelaars, T. (eds.) ECCV 2014. LNCS, vol. 8693, pp. 725–739. Springer, Cham (2014). https://doi.org/10.1007/978-3-319-10602-1_47
27. Carlinet, E., Géraud, T.: A comparative review of component tree computation algorithms. IEEE Trans. Image Process. **23**(9), 3885–3895 (2014)
28. Géraud, T., Carlinet, E., Crozet, S., Najman, L.: A quasi-linear algorithm to compute the tree of shapes of nD images. In: Hendriks, C.L., Borgefors, G., Strand, R. (eds.) ISMM 2013. LNCS, vol. 7883, pp. 98–110. Springer, Heidelberg (2013). https://doi.org/10.1007/978-3-642-38294-9_9
29. Stamatopoulos, N., Gatos, B., Kesidis, A.: Automatic borders detection of camera document images. In: 2nd International Workshop on Camera-Based Document Analysis and Recognition, Curitiba, Brazil (2007)
30. Gatos, B., Pratikakis, I., Perantonis, S.J.: Adaptive degraded document image binarization. Pattern Recogn. **39**(3), 317–327 (2006)
31. Chang, F., Chen, C.-J., Lu, C.J.: A linear-time component-labeling algorithm using contour tracing technique. Comput. Vis. Image Underst. **93**(2), 206–220 (2004)
32. Zhang, Z., He. L.-W.: Whiteboard Scanning and Image Enhancement (2016)
33. Skoryukina, N., et al.: Real-time rectangular document detection on mobile devices. In: Seventh International Conference on Machine Vision (ICMV 2014), vol. 9445. International Society for Optics and Photonics (2015)
34. Duda, R.O., Hart, P.E.: Use of the Hough transformation to detect lines and curves in pictures. Commun. ACM **15**(1), 11–15 (1972)

35. Skoryukina, N., et al.: Document localization algorithms based on feature points and straight lines. In: Tenth International Conference on Machine Vision (ICMV2017), vol. 10696. International Society for Optics and Photonics (2018)

36. Zhu, A., Zhang, C., Li, Z., Xiong, S.: Coarse-to-fine document localization in natural scene image with regional attention and recursive corner refinement. Int. J. Doc. Anal. Recog. **22**(3), 351–360 (2019). https://doi.org/10.1007/s10032-019-00341-0

37. Javed, K., Shafait, F.: Real-time document localization in natural images by recursive application of a CNN. In: 2017 14th IAPR International Conference on Document Analysis and Recognition (CDAR). vol. 1. IEEE (2017)

38. Awal, A.M., et al.: Complex document classification and localization application on identity document images. In: 2017 14th IAPR International Conference on Document Analysis and Recognition (ICDAR). vol. 1. IEEE (2017)

39. Baniadamdizaj, S., Soheili, M., Mansouri, A., et al.: Document localization in images taken by smartphones using a fully convolutional neural network, 04 October 2021, PREPRINT (Version 1). Research Square [https://doi.org/10.21203/rs.3.rs-952656/v1]

40. Chen, L.-C., Zhu, Y., Papandreou, G., Schroff, F., Adam, H.: Encoder-decoder with Atrous separable convolution for semantic image Segmentation. In: Ferrari, V., Hebert, M., Sminchisescu, C., Weiss, Y. (eds.) ECCV 2018. LNCS, vol. 11211, pp. 833–851. Springer, Cham (2018). https://doi.org/10.1007/978-3-030-01234-2_49

41. He, K., et al.: Spatial pyramid pooling in deep convolutional networks for visual recognition. IEEE Trans. Pattern Analy. Mach. Intell. **37**(9), 1904–1916 (2015)

42. Rahman, M.A., Wang, Y.: Optimizing intersection-over-union in deep neural networks for image segmentation. In: Bebis, G., et al. (eds.) ISVC 2016. LNCS, vol. 10072, pp. 234–244. Springer, Cham (2016). https://doi.org/10.1007/978-3-319-50835-1_22

43. Rezatofighi, H., et al.: Generalized intersection over union: a metric and a loss for bounding box regression. In: Proceedings of the IEEE Conference on Computer Vision and Pattern Recognition (2019)

Performance Comparison of Deep Residual Networks-Based Super Resolution Algorithms Using Thermal Images: Case Study of Crowd Counting

Syed Zeeshan Rizvi[(⊠)], Muhammad Umar Farooq, and Rana Hammad Raza

Electronics and Power Engineering Department, Pakistan Navy Engineering College (PNEC), National University of Sciences and Technology (NUST), Karachi, Pakistan
szrizvi.beee15pnec@student.nust.edu.pk,
{umar.farooq,hammad}@pnec.nust.edu.pk

Abstract. Humans are able to perceive objects only in the visible spectrum range which limits the perception abilities in poor weather or low illumination conditions. The limitations are usually handled through technological advancements in thermographic imaging. However, thermal cameras have poor spatial resolutions compared to RGB cameras. Super-resolution (SR) techniques are commonly used to improve the overall quality of low-resolution images. There has been a major shift of research among the Computer Vision researchers towards SR techniques particularly aimed for thermal images. This paper analyzes the performance of three deep learning-based state-of-the-art SR algorithms namely Enhanced Deep Super Resolution (EDSR), Residual Channel Attention Network (RCAN) and Residual Dense Network (RDN) on thermal images. The algorithms were trained from scratch for different upscaling factors of ×2 and ×4. The dataset was generated from two different thermal imaging sequences of BU-TIV benchmark. The sequences contain both sparse and highly dense type of crowds with a far field camera view. The trained models were then used to super-resolve unseen test images. The quantitative analysis of the test images was performed using common image quality metrics such as PSNR, SSIM and LPIPS, while qualitative analysis was provided by evaluating effectiveness of the algorithms for crowd counting application. After only 54 and 51 epochs of RCAN and RDN respectively, both approaches were able to output average scores of 37.878, 0.986, 0.0098 and 30.175, 0.945, 0.0636 for PSNR, SSIM and LPIPS respectively. The EDSR algorithm took the least computation time during both training and testing because of its simple architecture. This research proves that a reasonable accuracy can be achieved with fewer training epochs when an application-specific dataset is carefully selected.

Keywords: Crowd counting · Super resolution · Residual networks · EDSR · RCAN · RDN · Thermal imagery · PSNR · SSIM · LPIPS

© The Author(s) 2022
C. Biele et al. (Eds.): MIDI 2021, LNNS 440, pp. 75–87, 2022.
https://doi.org/10.1007/978-3-031-11432-8_7

1 Introduction

A normal human perception is limited to visible light spectrum only which ranges from ~380 nm to ~700 nm wavelength on the electromagnetic spectrum [1]. The visible cameras are designed to utilize the same wavelength range and replicate human vision by capturing RGB wavelengths for color representation. However, like a human eye, these systems are also affected by poor weather (such as fog, smoke, haze or storms etc.) and low illumination conditions. This limits their utilization to applications with only daytime scenarios. Beyond the visible spectrum is the infrared region which cannot be seen by humans. With the advancements in infrared/thermal imaging technologies, humans have extended their range of vision. These technologies enable the vision in most challenging situations such as low-visibility due to extreme weather conditions or low illumination etc. [2]. This is made possible because these cameras are essentially the heat sensors which can capture heat signatures from different objects. However, the spatial resolutions of these imaging technologies are relatively lower as compared to the visible cameras. Higher resolution is particularly important because it enables the capturing of small details in any image. The spatial resolution can be increased by use of a high-end camera, but that makes it a costly proposition. Many researchers use Image Super-resolution (SR) technique to reconstruct high resolution images from low resolution input images. SR is used to predict and fill details in low resolution images such that the output gives an image of a higher resolution. The resolution of the input image is increased based on the scaling factor used for super resolving the image.

SR is widely used in many computer vision applications including surveillance and security, small-objects detection and tracking to medical imaging. Most of the research regarding Image SR is focused towards images captured using visible cameras. However, surveillance environments are now commonly monitored using infrared/thermal cameras. The Computer Vision researchers are increasingly showing research interests in use of thermal images for a variety of applications [3–5]. Similarly, the same research trend is being noticed in SR applications using thermal images [6–9]. Rivadeneira et al. proposed a Convolutional Neural Network (CNN) based approach to compare performance of Image SR using both thermal and visible images [10]. The experimentation proved that the network trained with thermal images performed better than the latter.

This research work focuses on the use of deep Residual Network (ResNet) to perform SR of thermal images. ResNets have contributed significantly towards solving various Image SR issues. As presented in [11], a simple modification in traditional Convolutional Neural Networks (CNN) allows for bigger networks to be trained with an increased accuracy. The paper provides performance evaluation of some of the most popular ResNet architectures on the SR of thermal images. V. Chudasama et al. [12] provided a detailed comparison using Thermal Image Super Resolution (TISR) challenge dataset. The paper presented results of different state-of-the-art algorithms trained on this challenging dataset. Detailed analysis of large image datasets using state-of-the-art algorithms is though computationally expensive as it requires higher number of training epochs.

This research work specifically focuses on the application of crowd counting based on super-resolved thermal images using Enhanced Deep Super Resolution networks (EDSR), Residual Channel Attention Networks (RCAN) and Residual Dense Networks

(RDN) algorithms. The algorithms were trained from scratch using fixed camera views obtained from two different video sequences (suitable for crowd counting application) of BU-TIV benchmark dataset. The super-resolved image outputs generated by these algorithms were then used to count the number of persons using pretrained model weights. To obtain ground truth of the person count, the total number of persons were also predicted on the original ground truth image using these weights. The main contributions of this paper are highlighted as follows:

- A detailed comparative analysis of three most popular ResNet-based architectures i.e., EDSR, RCAN and RDN for thermal images SR to analyze crowd with inexpensive training dynamics. The crowd counting is performed on both sparse and highly dense static and dynamic nature crowd with added complexities due to far field camera viewing angles.
- Selection of a suitable application-specific dataset with fixed camera views for thermal images SR analysis in crowd counting applications. The sub-dataset is carefully selected to include both near and far field sparse and highly dense crowds for better visualization and understanding.

The rest of the paper is organized as followed. Section 2 provides a brief overview of the related research. Section 3 presents the working methodology of compared algorithms, and provides details regarding the implementation of this research work. Section 4 discusses the results obtained from the experimentation. Finally, Sect. 5 concludes the paper and provides a future research direction.

2 Related Work

SR of thermal images has gained much of the interest among researchers working in this area. One of the earliest approaches to SR of thermal images made use of the Huber Total Variation (HTV) approach which employed Huber norm with bilateral Total Variation (TV) regularization [13]. Chen et al. proposed the use of visible camera to guide SR of thermal images [14]. The approach was tested on their dataset and showed a reasonable performance while also avoiding traditional over-texture problem. Hans et al. proposed an SR algorithm for thermal images based on sparse representation [15]. Their results showed a good performance of image reconstruction without introducing major counterfeit artifacts. Cascarano et al. proposed an SR algorithm which can handle both single and multiple images [16]. The algorithm was tested on aerial and terrestrial thermal images and showed a good performance.

The very first deep learning implementation for Image SR was presented by Dong et al. [17] in 2014, with the introduction of SRCNN. In 2016, residual networks were introduced by He et al. for image recognition [11]. Architectures based on residual networks were then explored for single image SR. This helped make significant advancements in this area. In [18], SRResNet was introduced which was a 16 blocks deep residual network. Improvements were made in the SRResNet architecture by Lim et al. [19] with the introduction of EDSR. These networks became the backbone of major future research work in the domain of single image SR using residual networks. A recent approach proposed in [20] was built on residual blocks as base units. The architecture was used to generate super-resolved images in ×2, ×3 and ×4 scales and showed a good generalization capacity.

As the SR of thermal images is a relatively new research area, there is a need to explore this area in detail and build motivation for future research using specific applications. Therefore, this research gives an overview of application-specific SR algorithms implemented on thermal images.

3   Working Methodology of the Proposed Approach

3.1   Selected Algorithms

Three different deep learning-based SR algorithms were used in this study for comparative analysis. All algorithms are built on ResNet-based architectures. As discussed in Sect. 1, ResNets are a special case of CNN where small modifications in traditional CNN networks such as addition of skip connections enable the training of larger networks. Surprisingly, the resultant larger network does not cause performance degradation as in the previous CNN networks and generates even better accuracies, which makes it a perfect candidate for thermal images SR.

SR algorithms used in this research include EDSR, RCAN and RDN. The EDSR network [19] is inspired from SRResNet, the first ever ResNet used for Image SR. It removed batch normalization layer in SRResNet which improved results. The RCAN [21] uses Residual in Residual network which allows training of very deep CNNs for Image SR with major performance improvements. The RDN [22] introduced Residual Dense Blocks (RDB) which extract local features using dense convolutional layers.

Similarly, Focal Inverse Distance Transform (FIDT) maps were used for localization and counting of crowd [23]. These maps accurately localize the crowd without head overlaps even in highly dense environments.

3.2   Dataset

The thermal images used to conduct this research work were extracted from video sequences of BU-TIV dataset [24]. Only two video sequences i.e., Marathon-2 and Marathon-4 were suitable for crowd counting application. Both sequences provided a good view of cameras fixed at an elevated platform to address the problem of crowd counting. Figure 1 shows example frames from Marathon-2 and Marathon-4 videos. Both video sequences have a far field view which adds complexities in crowd estimation process. Videos are selected to include both sparse and highly dense crowd environments. The crowd has both static and dynamic motion features in multiple directions. The skewed camera angles and fewer number of pixels per head further make these video sequences challenging for crowd analysis. Both selected video sequences are perfect choices for crowd analysis and provide opportunities to explore performance of SR with given challenging attributes.

The data was recorded on FLIR SC8000 cameras. A total of 3555 frames were extracted from these video sequences and then reshaped into 512 × 256 resolution. These were used as ground truth images. To construct low-resolution images for ×2 and ×4 factors, ground truth images were down sampled two and four times into images with 256 × 128 and 128 × 64 resolutions respectively. All images from both sequences were randomly shuffled to improve generalizability of the model. Dataset was then split into train, validation, and test sets with ratios of 80:10:10 respectively.

3.3   Evaluation Setup

The training was done using NVIDIA's GeForce GTX 1080Ti. The deep learning framework used was PyTorch. The training parameters were kept same for all SR algorithms. Learning rate was fixed at 0.0001, ADAM was used as an optimizer function with $\beta 1$, $\beta 2$ and ϵ set at 0.9, 0.999 and 10^{-8} respectively. The networks were trained using L1 loss function. All networks were trained with 16 residual blocks and 64 feature maps to keep the comparison fair. Training was continued until the L1 loss reached a numerical value of 1.5 for ×2 upscaling, and 3.0 for ×4 upscaling. The threshold values were selected based on the observation that the loss graphs started to plateau around these values. The weights were saved as soon as L1 loss reached the threshold value. These weights were then used to super-resolve the 355 test images by factors of ×2 and ×4. The average Peak Signal-to-Noise Ratio (PSNR), Structural Similarity Index Measure (SSIM) and Learned Perceptual Image Patch Similarity (LPIPS) values were calculated from test images. PSNR provides a ratio between maximum power of an image and power of corrupting noise that affects its quality. SSIM calculates how similar two images are by comparing luminance, contrast, and structure between two images. LPIPS evaluates distance between two image patches.

Crowd counting was done on super-resolved test images through FIDT maps. The code was run using PyTorch backend. Pre-trained weights from University of Central Florida - Qatar National Research Fund (UCF-QNRF) dataset [25] were used for this crowd counting estimation.

(a)   (b)

Fig. 1. Views from BU-TIV dataset (a) Marathon-2 (b) Marathon-4

4 Results and Discussion

The training results obtained for ×2 upscaling factor are shown in Fig. 2. The L1 loss for RCAN reached the threshold value of 1.5 in the least number of epochs i.e., 72. RDN took 94 epochs to reach the value, whereas the training for EDSR had to be early stopped at 199 epochs as the graph had plateaued. For ×4 upscaling factor, training was continued until the L1 loss value crossed the numerical value of 3.0. The training results are displayed in Fig. 3. In this case, the L1 loss for RDN was able to reach threshold value in only 51 epochs. For RCAN, it took 54 epochs. The graph for L1 loss of EDSR had started to plateau in this case too, which is why training was early stopped at 300 epochs.

Weights obtained from the reported epochs were used for the testing phase. 355 unseen images were used in the testing phase. The test results are displayed in Table 1. It can be clearly observed that reasonable scores are achieved by all algorithms even when the maximum training epochs were only 300. An SSIM score close to 1.0, LPIPS score close to 0.0 and a high PSNR value represents images with a high level of structural similarity and near to being highly identical. The visual comparison of results on test images for ×2 and ×4 upscaling factors are displayed in Fig. 4 and Fig. 5 respectively. The frames are evaluated using the PSNR, SSIM and LPIPS metrics. It can be observed that the algorithms have accurately predicted static features in the video frames, e.g., the parked cars and the road. The slight difference in scores observable in the results is because of the dynamic features, e.g., moving cars and pedestrians. However, all algorithms showed improvements in the results as compared to the bicubic interpolated versions of the same images. The table also shows runtime analysis of each algorithm for 355 test images. Generally, the performance of EDSR was not as good as RCAN and RDN, but it has a considerably shorter execution time per step because of the simpler architecture. It was observed that the time taken with ×2 upscaling factor was more than the time taken for ×4 upscaling factor using RCAN and RDN. This is because the input images for ×2 upscaling factor are of a greater resolution than that of ×4 upscaling factor, as previously discussed in Sect. 3. It was also observed that application-specific datasets with fixed camera views are computationally efficient and generate robust detections with considerably few epochs.

Crowd counting using FIDT maps was done on all super-resolved images obtained in the testing phase by each method. To establish ground truth, the pretrained weights were used for crowd counting on the ground truth images. The Mean Absolute Error (MAE) and Root Mean Square Error (RMSE) scores obtained by performing counting through bicubic up-sampling and selected SR methods are presented in Table 2. RCAN performed better than all other selected SR algorithms using both ×2 and ×4 upscaling factors. A visual comparison of crowd count between ground truth and the obtained results is also provided in Fig. 6 with both ×2 and ×4 upscaling factors.

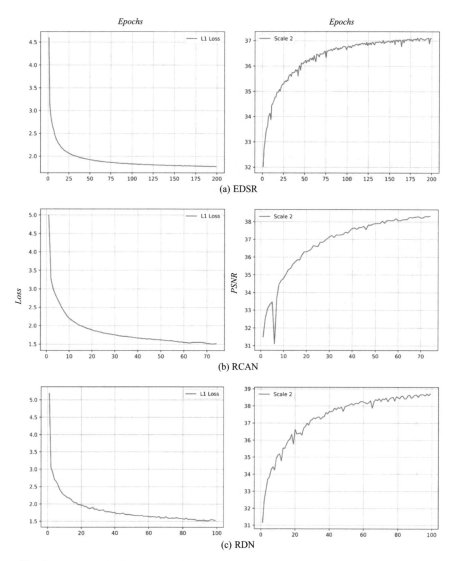

Fig. 2. L1 loss and PSNR scores obtained during training of SR algorithms for ×2 factor

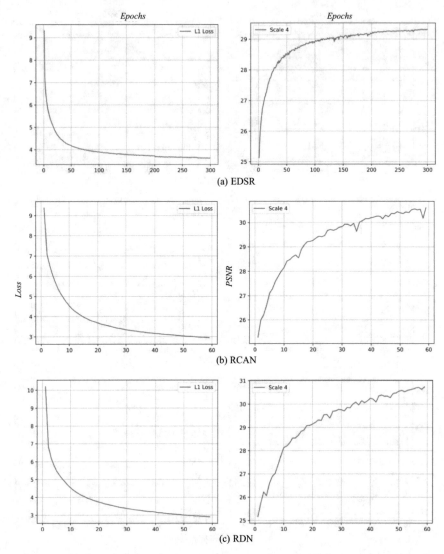

Fig. 3. L1 loss and PSNR scores obtained during training of SR algorithms for ×4 factor

Table 1. Quantitative comparison of selected SR algorithms on test data with $\times 2$ and $\times 4$ upscaling factors in terms of PSNR, SSIM and LPIPS metrics

Method	Scale	Time (s)	PSNR (dB)	SSIM	LPIPS
EDSR	$\times 2$	11.41	36.735	0.982	0.0171
	$\times 4$	11.46	28.933	0.926	0.0858
RCAN	$\times 2$	44.72	37.878	0.986	0.0098
	$\times 4$	30.72	30.102	0.945	0.0628
RDN	$\times 2$	54.95	38.227	0.986	0.0102
	$\times 4$	16.95	30.175	0.945	0.0636

Table 2. Performance evaluation of obtained crowd counting results with bicubic up-sampling and selected SR algorithms

Method	MAE		RMSE	
	$\times 2$	$\times 4$	$\times 2$	$\times 4$
Bicubic	24.78	40.09	37.09	68.49
EDSR	16.87	34.47	30.01	58.47
RCAN	**15.89**	**22.66**	**26.25**	**35.42**
RDN	17.26	26.40	28.04	39.78

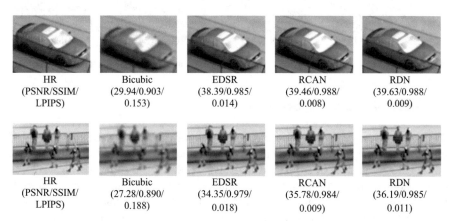

HR (PSNR/SSIM/ LPIPS)	Bicubic (29.94/0.903/ 0.153)	EDSR (38.39/0.985/ 0.014)	RCAN (39.46/0.988/ 0.008)	RDN (39.63/0.988/ 0.009)

HR (PSNR/SSIM/ LPIPS)	Bicubic (27.28/0.890/ 0.188)	EDSR (34.35/0.979/ 0.018)	RCAN (35.78/0.984/ 0.009)	RDN (36.19/0.985/ 0.011)

Fig. 4. Visual comparison of $\times 2$ SR results for both static and dynamic features

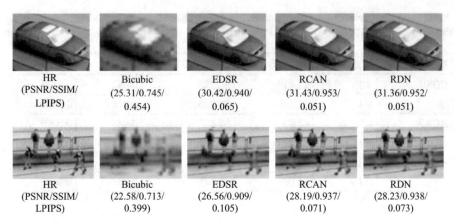

Fig. 5. Visual comparison of ×4 SR results for both static and dynamic features

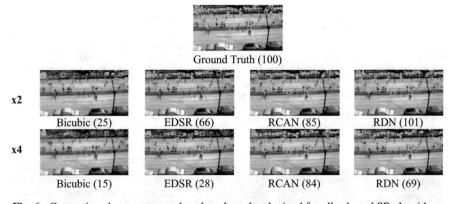

Fig. 6. Comparison between ground truth and results obtained for all selected SR algorithms

5   Conclusion and Future Work

This paper investigated the performance of state-of-the-art ResNet-based Image SR algorithms, namely EDSR, RCAN and RDN. The images were super-resolved ×2 and ×4 on video sequences of BU-TIV dataset. PSNR, SSIM and LPIPS scores were used as evaluation metrics to compare performance of each algorithm. As compared to the bicubic interpolated versions, all selected SR algorithms were able to generate good results due to their ResNet-based architectures which are proven to have good accuracies with deeper layers. With careful selection of a dataset with sufficient number of images for training, the models were able to perform good even with fewer epochs, the maximum of which were 300 epochs used by EDSR for ×4 up-scaling factor. The paper also provided a qualitative analysis by observing performance on crowd counting application in both sparse and highly dense crowd environments. RCAN outperformed other SR algorithms by achieving minimum MAE and RMSE values.

As a future work, a similar analysis can be performed with Image SR algorithms based on Generative Adversarial Networks (GANs). Furthermore, a completely new architecture can also be designed particularly tailored for crowd counting application using SR on thermal images. Similarly, different multi-image SR methods can also be explored for performance comparison with single image SR algorithms.

Acknowledgement. We acknowledge support from National Center of Big Data and Cloud Computing (NCBC) and Higher Education Commission (HEC) of Pakistan for conducting this research.

Data Availability Statement.   All related data including dataset, trained model weights and high-resolution results are placed on following Google Drive link: https://drive.google.com/drive/fol ders/1LNLVVNCRDRIP__lN4DJSjxmws15BWcpw?usp=sharing (Last accessed on 15 Nov 21).

References

1. NASA. Visible Light | Science Mission Directorate. https://science.nasa.gov/ems/09_visibl elight. Accessed 15 Nov 2021
2. Kristoffersen, M., Dueholm, J., Gade, R., Moeslund, T.: Pedestrian counting with occlusion handling using stereo thermal cameras. Sensors **16**(1), 62 (2016). https://doi.org/10.3390/s16 010062.10.1007/s00521-021-05973-0
3. Fernandes, S.L., Rajinikanth, V., Kadry, S.: A hybrid framework to evaluate breast abnormality using infrared thermal images. IEEE Consum. Electron. Mag **8**(5), 31–36 (2019). https://doi. org/10.1109/mce.2019.2923926
4. Ghose, D., et al.: Pedestrian detection in thermal images using saliency maps: In: Proceedings of the IEEE/CVF Conference on Computer Vision and Pattern Recognition Workshop (2019)
5. Zeng, X., Miao, Y., Ubaid, S., Gao, X., Zhuang, S.: Detection and classification of bruises of pears based on thermal images. Postharv. Biol. Technol. **161**, 111090 (2020). https://doi.org/ 10.1016/j.postharvbio.2019.111090
6. Patel, H., et al.: ThermISRnet: an efficient thermal image super-resolution network. Opt. Eng. **60**(07) (2020). https://doi.org/10.1117/1.oe.60.7.073101.10.1038/s41598-020-77979-y
7. Ahmadi, S., et al.: Laser excited super resolution thermal imaging for nondestructive inspec- tion of internal defects. Sci. Rep. **10**(1) (2020). https://doi.org/10.1038/s41598-020-779 79-y
8. Kuni Zoetgnande, Y.W., Dillenseger, J.-L., Alirezaie, J.: Edge focused super-resolution of thermal images. In: 2019 International Joint Conference on Neural Networks (IJCNN) (2019). https://doi.org/10.1109/ijcnn.2019.8852320.10.3390/s21041265
9. Raimundo, J., Lopez-Cuervo Medina, S., Prieto, J.F., Aguirre de Mata, J.: Super resolu- tion infrared thermal imaging using Pansharpening algorithms: quantitative assessment and application to UAV thermal imaging. Sensors **21**(4), 1265 (2020). https://doi.org/10.3390/s21 041265
10. Rivadeneira, R.E., Suárez, P.L., Sappa, A.D., Vintimilla, B.X.: Thermal image SuperResolu- tion through deep convolutional neural network. In: Karray, F., Campilho, A., Yu, A. (eds.) ICIAR 2019. LNCS, vol. 11663, pp. 417–426. Springer, Cham (2019). https://doi.org/10. 1007/978-3-030-27272-2_37

11. He, K., Zhang, X., Ren, S., Sun, J.: Deep residual learning for image recognition. In 2016 IEEE Conference on Computer Vision and Pattern Recognition (CVPR) (2016). https://doi.org/10.1109/cvpr.2016.90

12. Chudasama, V., et al.: Therisurnet-a computationally efficient thermal image super-resolution network. In: Proceedings of the IEEE/CVF Conference on Computer Vision and Pattern Recognition Workshops (2020)

13. Panagiotopoulou, A., Anastassopoulos, A.: Super-resolution reconstruction of thermal infrared images. In: Proceedings of the 4th WSEAS International Conference on REMOTE SENSING (2008)

14. Chen, X., Zhai, G., Wang, J., Hu, C., Chen, Y.: Color guided thermal image super resolution. In: 2016 Visual Communications and Image Processing (VCIP) (2016). https://doi.org/10.1109/vcip.2016.7805509

15. Jino Hans, W., Venkateswaran, N.: An efficient super-resolution algorithm for IR thermal images based on sparse representation. In: Proceedings of the 2015 Asia International Conference on Quantitative InfraRed Thermography (2015). https://doi.org/10.21611/qirt.2015.0092.10.3390/rs12101642

16. Cascarano, P., et al.: Super-resolution of thermal images using an automatic total variation based method. Remote Sens. **12**(10), 1642 (2020). https://doi.org/10.3390/rs12101642

17. Dong, C., Loy, C.C., He, K., Tang, X.: Learning a deep convolutional network for image super-resolution. In: Fleet, D., Pajdla, T., Schiele, B., Tuytelaars, T. (eds.) ECCV 2014. LNCS, vol. 8692, pp. 184–199. Springer, Cham (2014). https://doi.org/10.1007/978-3-319-10593-2_13

18. Ledig, C., et al.: Photo-realistic single image super-resolution using a generative adversarial network. In: Proceedings of the IEEE Conference on Computer Vision and Pattern Recognition (2017)

19. Lim, B., et al.: Enhanced deep residual networks for single image super-resolution. In: Proceedings of the IEEE Conference on Computer Vision and Pattern Recognition Workshops (2017)

20. Kansal, P., Nathan, S.: A multi-level supervision model: a novel approach for thermal image super resolution. In: Proceedings of the IEEE/CVF Conference on Computer Vision and Pattern Recognition Workshop. (2020)

21. Zhang, Y., Li, K., Li, K., Wang, L., Zhong, B., Fu, Y.: Image super-resolution using very deep residual channel attention networks. In: Ferrari, V., Hebert, M., Sminchisescu, C., Weiss, Y. (eds.) ECCV 2018. LNCS, vol. 11211, pp. 294–310. Springer, Cham (2018). https://doi.org/10.1007/978-3-030-01234-2_18

22. Zhang, Y., Tian, Y., Kong, Y.; Zhong, B., Fu, Y: Residual dense network for image super-resolution. In: 2018 IEEE/CVF Conference on Computer Vision and Pattern Recognition (2018). https://doi.org/10.1109/cvpr.2018.00262

23. Liang, D., Xu, W., Zhu, Y., Zhou, Y.: Focal inverse distance transform maps for crowd localization and counting in dense crowd. arXiv:2102.07925 [cs] (2021)

24. Wu, Z., Fuller, N., Theriault, D., Betke, M.: A thermal infrared video benchmark for visual analysis. In: 2014 IEEE Conference on Computer Vision and Pattern Recognition Workshops (2014). https://doi.org/10.1109/cvprw.2014.39

25. Idrees, H., et al.: Composition loss for counting, density map estimation and localization in dense crowds. arXiv:1808.01050 [cs] (2018)

Normalization and Bias in Time Series Data

Aishwarya Asesh[(⊠)]

Adobe, Mountain View, USA
a.asesh@gmail.com

Abstract. Data normalization is an important preprocessing step in data mining and Machine Learning (ML) technique. Finding an acceptable approach to deal with time series normalization, on the other hand, is not an easy process. This is because most standard normalizing approaches rely on assumptions that aren't true for the vast majority of time series. The first is that all time series are stationary, which means that their statistical characteristics, such as mean and standard deviation, do not vary over time. The time series volatility is assumed to be uniform in the second assumption. These concerns are not addressed by any of the approaches currently accessible in the literature. This research provides theoretical and experimental evidence, that normalizing time series data, can prove to be of utmost value by trimming non necessary data points and achieving minimum information loss, by using the concept of Minimal Time Series Representation (MTSR).

Keywords: Time series · Normalization · Classification · Time series length · Minimal time series representation

1 Introduction

The University of California, Riverside (UCR) Time Series Classification Archive [1], has grown into a valuable resource for the time series data mining community, with over a thousand articles citing at least one data set from the repository. While the classification accuracy demonstrated by predictive models on UCR data is undeniable, it is critical to look into the impact of data normalization approaches on classification accuracy. Because data normalization procedures are known to have a substantial influence on prediction accuracy for many classifiers, a knowledge of the impact of UCR's approaches is required to validate the accuracy of classification models.

Due to the bias incorporated into time series classification approaches, created and evaluated on a single benchmark dataset, as discussed by Keogh and Kasetty [3], there is a clear need for broader testing on real-world data. However, it is important to validate if it's not just the data that causes methods to become over-trained, but also the normalization that goes into producing such datasets.

With the aid of raw unprocessed and non-normalized UCR data provided by Geoff Web, Anthony Bagnall and Eamonn Keogh, this research study focuses

© The Author(s) 2022
C. Biele et al. (Eds.): MIDI 2021, LNNS 440, pp. 88–97, 2022.
https://doi.org/10.1007/978-3-031-11432-8_8

on normalization techniques and understanding the influence of normalization approaches on classifier (and regressor) accuracy.

2   Normalization and Time Series Data

Normalization methods are recognized to have a significant impact on classification accuracy in multivariate data sets. When data from two distributions with vastly different means and variances exist, normalization becomes critical in guaranteeing that each variable does not bias prediction. This may be less essential in univariate datasets [4,5]. Despite this, the multidimensional issue space becomes simpler to train in a variety of predictive models, including neural networks and support vector classifiers, and a number of mathematical functions rely on normalized data. The choice of activation functions in neural networks is greatly influenced by this fact, with sigmoid activations becoming essentially useless until input is in the 0–1 range. If the hyperplanes used in class separation can be fitted most precisely, support vector machines will require a standardised problem space. While this is a more complicated topic in and of itself, this aforementioned research concentrates on time series length standardisation.

While each data set in the UCR Time Series Data Archive has a drastically varied duration, it is crucial to identify the influence this has on categorization. The varying rates at which events occur throughout a number of occurrences is one reason why Dynamic Time Wrapping (DTW) distance measurements are so useful. When an event occurs in a specific length of time, it is possible that the same event will occur in a greater time frame in another occurrence, euclidean distance measurements will not match in these cases [6,7]. Although altering the time series length will not address this problem, it is critical in evaluating how much information is necessary in a specific time series to get best outcomes. The remainder of the raw time series is not necessary, if using a very short time series length is able too achieve high classification accuracy. This concept is comparable to early detection, which is a distinct field of time series classification. One can regulate which portion of the information and how much of it classifiers may utilize for prediction by altering the length of time series, both where they begin and where they stop [8,9]. While having more data provides for higher prediction accuracy, it also introduces noise, and the larger the data, the longer it takes to classify it. The information gain/loss, as well as the slow-down/speed-up associated with it, may be understood through changes in time series length.

3   Experiments

The data from the UCR Time Series Classification Archive is not only standardized using z-score normalization, but it is also divided into training and testing subsets. The raw data required to create the UCR datasets includes CricketX, CricketY, CricketZ, GesturesX, GesturesY, GesturesZ, and Wafer. These datasets will be referred to as Cricket, Gestures, and Wafer. Two techniques are used to assess the accuracy of categorization on this data.

The first involves creating a distribution of classification accuracy for a particular data set using random train/test splits of the same size. The second technique is used to see whether there are any discrepancies between the raw data and the data in the UCR repository. To evaluate if there are any major discrepancies between the two datasets, each time series in the raw data is matched with its closest matching time series in the UCR repository data.

3.1 Normalization

Scalar Normalization. The data is normalized using the z-score method. The two most prevalent scale normalizing approaches, z-score normalization and min-max normalization, are used. When the data corresponds to a normal distribution, Z-score normalization is the most frequent and most representational of the original raw data. The process of Z-score normalization entails turning each data point into a positive or negative number that represents how many standard deviations the data point is from the mean.

Min-max normalization includes removing the minimum and dividing by the difference between the time series datapoint minimum and maximum values to transform data into values between 0 and 1. These normalization approaches are straightforward and widely used to eliminate bias in variables with larger values when compared to data with lower values.

Time Series Length Normalization. The data is organized so that each time series in a dataset is the same length. There is no need for stringent normalization techniques with regard to time series length, as in datasets like Cricket, the lengths of time series have a relatively little deviation from the mean. Despite this, when compared to the run durations of the closest neighbor technique, time series length minimization gives a significant computational speedup. The impact of both increasing and lowering the length of time series is an important aspect. Both approaches entail shrinking/expanding a *n-length* time series to a *m-length* time series.

For a time series T of length n, the i-th data point in T is represented as T_i. T is converted to a time series of length m, which is denoted as S. The new time series S where each data point in S is as follows:

$$S_i = T_j \quad j = \lfloor n \times \frac{i}{m} \rfloor$$

Information loss is observed when m is less than n, the impact of which will be detailed in the coming sections.

3.2 Classification Techniques

Using a variety of approaches, the UCR Time Series Classification Archive defines the lowest classification error achievable.

1 1-Nearest Neighbor classifier - 1-NN Euclidean - This is the error produced by utilizing one nearest neighbor algorithm with a euclidean distance metric.

2 1-NN DTW Best Warping Window - In the NN-DTW classification technique, the warping window is a hyper-parameter that has been computed for each data set, coupled with the error attained with this ideal window.

3 1-NN DTW DTW DTW DTW DTW DTW No Warping Window - When the NN-DTW classifier has no warping window, this error occurs.

The correctness of normalized data can be determined by comparing these three approaches. Because of their strength in time series classification [2], the universality and simplicity of the algorithms, these three approaches are chosen. While there are alternative approaches that may yield more accurate classifications, using NN classifiers allows for a better understanding of the homogeneity and relatedness of the normalized data within a particular class.

4   Results and Discussion

While the datasets are extremely diverse in origin, there are obvious parallels in terms of classification accuracy for both Euclidean and DTW distance metrics (Fig. 1).

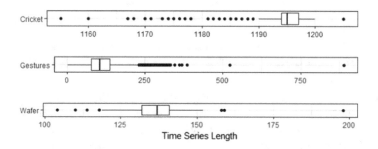

Fig. 1. The distribution of time series length for each of the three datasets.

As the time series length grows, the classification accuracy approaches a maximum, as seen in Fig. 2 and Fig. 3. However, the maximum classification accuracy is obtained at a relatively short time series length and remains almost constant as the duration grows.

The concept of a Minimal Time Series Representation (MTSR), which minimizes data information loss (along with maximizing classification accuracy) is established. It can be observed that there is a minimum information loss in the instance of *Cricket* at a time series length close to the beginning values of the time series, 1200. While this reduces information loss, one can obtain the same classification accuracy with a time series length of 40 and a slightly greater classification accuracy distribution with a time series length of 90. The primary advantage of an MTSR is that it allows for quicker classification owing to the lower computing burden of conducting Euclidean distance measurements on a smaller time series.

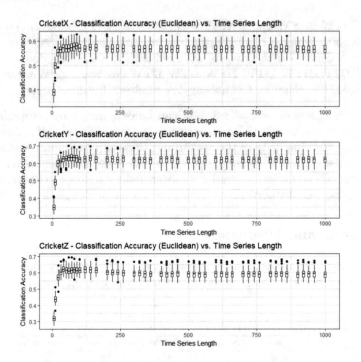

Fig. 2. The distribution of classification accuracy for the *Cricket* dataset as a function of time series length using the Euclidean distance measure in NN prediction.

Similar results are observed for *Wafer* dataset, as illustrated in Fig. 3, where the classification accuracy stays steady as the time series length increases. One can detect a clear maximum classification accuracy for a length of 30 in this dataset. With a time series length of 10–20, a similar accuracy can be achieved as with a length of 300 (Fig. 4).

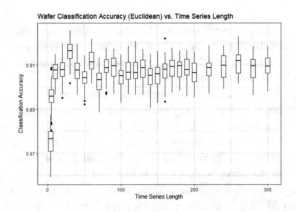

Fig. 3. The distribution of classification accuracy for the *Wafer* dataset as a function of time series length using the Euclidean distance measure in NN prediction.

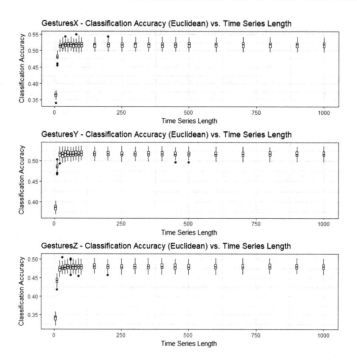

Fig. 4. The distribution of classification accuracy for the *Gestures* dataset as a function of time series length using the Euclidean distance measure in NN prediction.

The results and findings utilizing a DTW distance measure differ considerably from those obtained using Euclidean distance measurements. There is a definite maximum classification accuracy in each of the three datasets. With the following optimum time series lengths and classification accuracies for each dataset:

Dataset	Avg. accuracy	Optimal length	Optimal accuracy
CricketX	0.6299	100	0.7756
CricketY	0.7063	100	0.8353
CricketZ	0.6888	100	0.8061
GesturesX	0.5768	60	0.6459
GesturesY	0.5708	40	0.6498
GesturesZ	0.5293	50	0.5975
Wafer	0.9142	325	0.9214

There is a considerable absolute gain in classification accuracy in both the *Cricket* and *Gesture* datasets. While in case of *Wafer*, it appears to be merely a minor gain in classification accuracy, it is 0.77 standard deviations above the mean accuracy across all duration. This rise in accuracy over the mean is true for both *Cricket* and *Gesture*, with optimal time series length accuracy in *Cricket*

X, Y, and Z being 1.66, 1.36, and 1.24 standard deviations above the mean, respectively. *Gesture* X, Y, and Z have ideal time series length accuracy of 1.12, 1.23, and 1.15 standard deviations above the mean, respectively. These are all considerable gains over average accuracy for various time series lengths, but the most interesting finding is that there is an ideal time series length for DTW accuracy. DTW has an unambiguous maximum with regard to time series length, unlike Euclidean measures, which have a minimal representation of the time series that nevertheless maximizes classification accuracy.

Because of the single maximum, there is considerable symmetry, with lengths on each side of the optimal length resulting in the same accuracy. In *Cricket*, for example, time series lengths of 30 and 550 produce extremely equal classification accuracy, but picking the length of 30 would be optimum for speed of calculation.

There is no simple answer to determining a generic minimal time series length, as there is with practically all sample size questions. It is determined by the number of model parameters to be evaluated as well as the quantity of data randomness. With the number of parameters to be estimated and the level of noise in the data, the sample size required grows (Figs. 5 and 6).

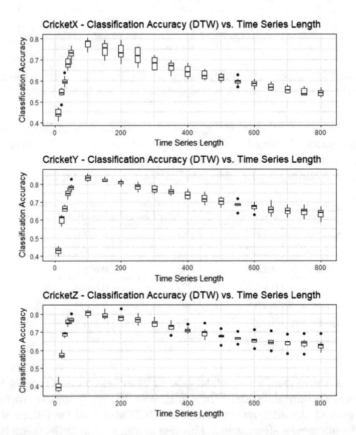

Fig. 5. The distribution of classification accuracy for the *Cricket* dataset as a function of time series length using the DTW distance measure in NN prediction.

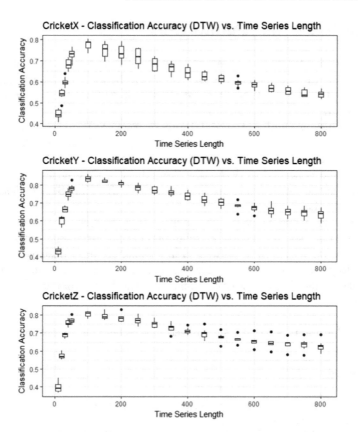

Fig. 6. The distribution of classification accuracy for the *Cricket* dataset as a function of time series length using the DTW distance measure in NN prediction.

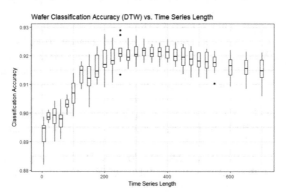

Fig. 7. The distribution of classification accuracy for the *Wafer* dataset as a function of time series length using the DTW distance measure in NN prediction.

5 Conclusions and Future Work

While acknowledged to be less accurate than DTW in time series classification, Euclidean distance measurements have shown to be more stable than DTW. In comparison, DTW has a time series length that is optimal for highest classification accuracy.

The findings are limited to the datasets mentioned. While results about time series length and normalization approaches apply to various datasets, the nature of the data matters when it comes to classification accuracy. As a result, additional research is needed to determine the impact of time series length normalization and scalar normalizing on data from the UCR Time Series Classification Archive and other sources.

In addition, more research into multivariate time series is necessary. The information loss associated with the reduction in time series length has an impact on classification accuracy, as addressed in this paper (both positively and negatively). More study is needed to develop more complex models for determining the smallest time series representations with the least amount of information loss (Fig. 7).

Acknowledgement. All that I am, or ever hope to be, I owe to my angel mother.

References

1. Chen, Y., et al.: The UCR Time Series Classification Archive (2015). www.cs.ucr.edu/~eamonn/time_series_data/
2. Petitjean, F., Forestier, G., Webb, G.I., Nicholson, A.E., Chen, Y., Keogh, E.: Faster and more accurate classification of time series by exploiting a novel dynamic time warping averaging algorithm. Knowl. Inf. Syst. **47**(1), 1–26 (2015). https://doi.org/10.1007/s10115-015-0878-8
3. Keogh, E., Kasetty, S.: On the need for time series data mining benchmarks: a survey and empirical demonstration. Data Min. Knowl. Discov. **7**(4), 349–371 (2003)
4. Deng, J., Chen, X., Jiang, R., Song, X., Tsang, I.W.: ST-Norm: spatial and temporal normalization for multi-variate time series forecasting. In: Proceedings of the 27th ACM SIGKDD Conference on Knowledge Discovery and Data Mining, pp. 269–278 (2021)
5. Dette, H., Kokot, K., Volgushev, S.: Testing relevant hypotheses in functional time series via self-normalization. J. R. Stat. Soc. Ser. B (Stat. Methodol.) **82**(3), 629–660 (2020)
6. Boubrahimi, S.F., Hamdi, S.M., Ma, R., Angryk, R.: On the mining of the minimal set of time series data Shapelets. In: 2020 IEEE International Conference on Big Data (Big Data), pp. 493–502. IEEE (2020)
7. Denison, R.N., Carrasco, M., Heeger, D.J.: A dynamic normalization model of temporal attention. Nat. Hum. Behav. 1–12 (2021)
8. Passalis, N., Tefas, A., Kanniainen, J., Gabbouj, M., Iosifidis, A.: Deep adaptive input normalization for time series forecasting. IEEE Trans. Neural Netw. Learn. Syst. **31**(9), 3760–3765 (2019)

9. Chonbodeechalermroong, A., Ratanamahatana, C.A.: Robust scale-invariant normalization and similarity measurement for time series data. In: Sieminski, A., Kozierkiewicz, A., Nunez, M., Ha, Q.T. (eds.) Modern Approaches for Intelligent Information and Database Systems. SCI, vol. 769, pp. 149–160. Springer, Cham (2018). https://doi.org/10.1007/978-3-319-76081-0_13

Deep Neural Networks Approach to Microbial Colony Detection—A Comparative Analysis

Sylwia Majchrowska[1,2(✉)], Jarosław Pawłowski[1,2], Natalia Czerep[1,3],
Aleksander Górecki[1,2], Jakub Kuciński[1,3], and Tomasz Golan[1]

[1] NeuroSYS, Rybacka 7, 53-656 Wrocław, Poland
`sylwia.majchrowska@pwr.edu.pl`
[2] Wroclaw University of Science and Technology,
Wybrzeże S. Wyspiańskiego 27, 50-372 Wroclaw, Poland
[3] University of Wroclaw, Fryderyka Joliot-Curie 15, 50-383 Wroclaw, Poland

Abstract. Counting microbial colonies is a fundamental task in microbiology and has many applications in numerous industry branches. Despite this, current studies towards automatic microbial counting using artificial intelligence are hardly comparable due to the lack of unified methodology and the availability of large datasets. The recently introduced AGAR dataset is the answer to the second need, but the research carried out is still not exhaustive. To tackle this problem, we compared the performance of three well-known deep learning approaches for object detection on the AGAR dataset, namely two-stage, one-stage, and transformer-based neural networks. The achieved results may serve as a benchmark for future experiments.

Keywords: Deep learning · Object detection · Microbial colony counting

1 Introduction

The ability to automatically and accurately detect, localize, and classify bacterial and fungal colonies grown on solid agar is of wide interest in microbiology, biochemistry, food industry, or medicine. An accurate and fast procedure for determining the number and type of microbial colonies grown on a Petri dish is crucial for economic reasons - industrial testing often relies on proper determination of colony forming units (CFUs). Conventionally, the analysis of samples is performed by trained professionals, even though it is a time-consuming and error-prone process. To avoid these issues, an automated methodology based on artificial intelligence can be applied.

A common way of counting and classifying objects using deep learning (DL) is to first detect them and then count the found instances distinguishing between different classes. We compared the results of microbe colony counting using

C. Biele et al. (Eds.): MIDI 2021, LNNS 440, pp. 98–106, 2022.
https://doi.org/10.1007/978-3-031-11432-8_9

selected detectors belonging to different classes of neural network architectures, namely two-stage, one-stage, and transformer-based models.

Our experiments were conducted on the Annotated Germs for Automated Recognition (AGAR) dataset with *higher-resolution* subset [10], which consists of around 7k annotated Petri dish photos with five microbial classes. This paper focuses on setting benchmarks for the detection and counting tasks using state-of-the-art (SoTA) models.

2 Object Detection

Object detection was approached using two major types of DL-based architectures, namely two-stage and one-stage models. **Two-stage** detectors find class-agnostic object proposals in the first stage, and in the second stage the proposed regions are assigned to the most likely class. They are characterised by high localization and classification accuracy. The largest group of two-stage models are Region Based Convolutional Neural Networks (R-CNN) [6], whose main idea is based on extracting region proposals from the image. Over the years, the networks from this family have undergone many modifications. In the case of *Faster R-CNN* [15] architecture, a Region Proposal Network (RPN) was used instead of the Selective Search algorithm. This allows for significant reduction of the model's inference time. In order to reduce issues with over-fitting during training, *Cascade R-CNN* [2] was introduced as multi-stage object detector, which consists of multiple connected detectors that are trained with increased intersection over union (IoU) thresholds. A year later, authors of *Libra R-CNN* [11] focused on balancing the training process by IoU-balanced sampling, balanced feature pyramid, and balanced L1 loss. In the following years, researchers used Faster R-CNN while replacing its backbone (CNN used as a feature extractor) with newer architectures. The most recent concept is *Composite Backbone Network V2* [8] (CBNetV2), which groups multiple pretrained backbones of the same kind for more efficient training.

On the other hand, the **one-stage** architectures are designed to directly predict classes and bounding box locations. For this reason, one-stage detectors are faster, but usually have relatively worse performance. Single-stage detection was popularized in DL mainly by You Only Look Once models (YOLO v1 [12], 9000 [13], v3 [14], v4 [1]), primarily developed by Joseph Redmon. Recently, the authors of *YOLOv4* [1] have enhanced the performance of YOLOv3 architecture using methods such as data augmentation, self-adversarial training, and class label smoothing, all of which improve detection results without degrading the inference speed. Moreover, the authors of *EfficientDet* [16] introduce changes which contribute to an increase in both accuracy and time performance of object detection. The main proposed changes include using weighted Bidirectional Feature Pyramid Network (BiFPN), compound scaling, and replacing the backbone network with EfficientNet [17]. EfficientNet as a backbone connects with the idea of scalability from compound scaling, which allows the model to be scaled to different sizes and to create a family of object detectors for specific uses.

Additionally, the **transformers** have recently become a next generation of neural networks for all computer vision applications, including object detection [3, 5, 19]. The interest in replacing CNN with transformers is mainly due to their efficient memory usage and excellent scalability to very large capacity networks and huge datasets. The parallelization of transformer processes is achieved by using an attention mechanism applied to images split into patches treated as tokens. The utilization of the transformer architecture to generate predictions of objects and their position in an image was first proposed in DEtection TRansformer (DETR) network [3]. The architecture uses a CNN backbone to learn feature maps, then feeds transformer layers. In comparison to DETR, *Deformable DETR* [19] network replaces self-attention in the encoder and cross-attention in the decoder with multi-scale deformable attention and cross-attention. Deformable attention modules only attend to a small set of key sampling points around a reference point which highly speeds up the training process. The recently introduced *Cross-Covariance Image Transformers* (XCiT) [5] concept is a new family of transformer models for image processing. The idea is to use a transformer-based neural network as a backbone for two-stage object detection networks. XCiT splits images into fixed size patches and reduces them into tokens with a greater number of features with the use of a few convolutional layers with Gaussian Error Linear Units (GELU) [7] in between. The idea behind the model is to replace self-attention with transposed attention (which is over feature maps instead of tokens).

3   AGAR Dataset

The AGAR dataset [10] contains images of microbial colonies on Petri dishes taken in two different environments, which produced *higher resolution* and *lower resolution* images. The differences are between the lighting conditions and apparatuses. *Higher resolution* images, which were used in our studies, can be divided into *bright*, *dark* and *vague* subgroups. On the other hand, considering the number of colonies, samples can be defined as *empty*, *countable* and *uncountable*. The dataset includes five classes, namely *E.coli*, *C.albicans*, *P.aeruginosa*, *S.aureus*, *B.subtilis*, while annotations are stored in json format with information about the number and type of microbe, environment and coordinates of bounding boxes.

In this paper, we present the results of experiments performed using a subset of the AGAR dataset, which consists of 6990 images in total. In our case only *higher resolution* (mainly 4000 × 4000 px), *dark* and *bright*, without *vague*, samples with *countable* number of colonies were chosen. Firstly, images were split into train and validation subsets (the same for each experiment), and then divided into 512 × 512 px patches as described in [10]. At the end—in the test stage—whole images from the validation subset of the Petri dish were used (for a detailed description of the procedure see Supplementary materials from [10]).

4   Benchmarking Methodology

We compared the performance of the selected models using several metrics: architecture type and size, inference time, and detection and counting accuracy.

During time measurements, the inference was executed on GeForce GTX 1080 Ti GPU using the same patch with 6 ground truth instances. The models were first loaded into memory, then inferred 100 times sequentially (ignoring the first 20 times for warming up) to calculate averaged time and its standard deviation for each model separately.

As to detection results, the detector performance was evaluated twofold - by measuring the effectiveness of detection and counting. As an evaluation metric for colony detection, we rely on the mean Average Precision (mAP), to be precise mAP@.5:.95, averaged over all 5 classes. The efficiency of colony counting was measured based on Mean Absolute Error (MAE), and Symmetric Mean Absolute Percentage Error (sMAPE).

With the growing popularity of DL, many open source software libraries implementing SoTA object detection algorithms emerge. Results provided for Faster R-CNN and Cascade R-CNN were taken from [10] for comparison purposes. Similarly, in our experiments we relied on MMDetection [4] framework (Libra R-CNN, CBNetV2, Deformable DETR, XCiT), Alexey Bochkovskiy's Darknet-based implementation of YOLOv4 [1], and Ross Wightman's PyTorch [18] reimplementation of official EfficientDet's TensorFlow implementation. To perform model training, we used the default parameters as for COCO dataset in the above-mentioned implementations. In all experiments, we used transfer learning technique based on pretrained backbones (ResNet-50, Double ResNet-50, XciT-T12) on ImageNet or whole architecture (YOLOv4, EfficientDet-D2, Deformable DETR) on MS COCO dataset. In the case of YOLOv4, we changed the input size to 512×512 px in order to match the size of the generated patches. We used pretrained backbones in all experiments. Traditional two- and one-stage networks were trained with Stochastic Gradient Descent (SGD) optimizer, as opposed to Transformer based architectures, where AdamW [9] optimizer was used. The values of initial learning rate vary between 10^{-3} and 10^{-5} for each model. All networks were trained until loss values saturated for the validation subset. We also chose commonly used augmentation strategies of selected models, like flips, crops, and resizes of images.

4.1 Results

Mean averaged precisions presented in Table 1 are averaged over all microbe classes. Calculated value of mAP@.5:.95 varies between 0.491 and 0.529. The most efficient results in terms of accuracy and inference speed were achieved for YOLOv4 architecture. On the other hand, transformer-based architectures present slightly worse performance. Some interesting cases were presented in Fig. 1. The selected image presents the same microbial species (*P. aeruginosa*), which forms two different sizes of colonies due to agar inhomogeneity, making detection even more challenging. Labeled small contamination is not perceived by all models (transformer based and EfficientDet-D2), and some of them (YOLOv4, Deformable DETR) also have problems with precise localization of blurred colonies. Two-stage detectors have a tendency to produce some excessive predictions.

Table 1. Benchmarks for tested models on the *higher-resolution* subset of AGAR dataset. The model size is given in terms of number of parameters (in millions). In case of XCiT model number of backbone's parameters is given in brackets.

Type	Model	Backbone	mAP .5:.95	Inference time (ms)	Size (M)
Two-stage	Faster R-CNN	ResNet-50	0.493	54.06 ± 1.72	42
	Cascade R-CNN	ResNet-50	0.516	76.31 ± 1.96	69
	Libra R-CNN	ResNet-50	0.499	33.34 ± 0.49	41
	CBNetv2	Double ResNet-50	0.502	43.79 ± 0.42	69
One-stage	YOLOv4	CSPDarknet53	0.529	17.46 ± 0.17	64
	EfficientDet-D2	EfficientNet-B2	0.512	45.59 ± 1.06	8
Transformer	Deformable DETR	ResNet-50	0.492	72.40 ± 0.65	40
Transformer backbone	Faster R-CNN	XCiT-T12	0.491	110.99 ± 3.85	25 (7)

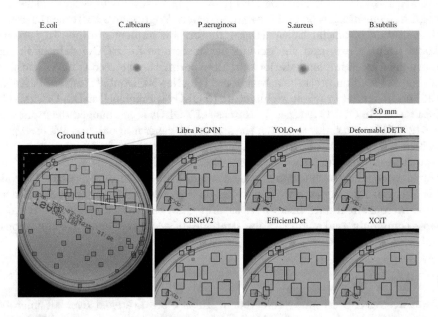

Fig. 1. Examples of images of microbial species and achieved predictions for a selected sample. Whole image of the Petri dish presents ground truth annotations, while the white dashed rectangle indicates the region chosen for visualisation of predicted results. Red rectangles mark *P. aeruginosa* species, the black one – *Contamination* (microorganism not intentionally cultured), green – *C. albicans*, navy – *S. aureus*.

The performance of the selected architectures for microbial counting is presented in both Table 2 and Fig. 2, while Table 3 shows all five microbial species separately. In general, all detectors perform better for microbes that form clearly visible, separate colonies. The biggest problem with locating individual colonies was observed for *P. aeruginosa*, where the tendency for aggregation and overlapping is the greatest. Overall, the best results were obtained for the YOLOv4 model, where the predicted count of microbial colonies is the closest to ground truth in range from 1 to 50 instances (see Fig. 2) – the most operable scope for industrial applications. The straight black lines represent a 10% error to highlight the acceptable error range indicated by microbiologists. The worst performance was observed for the EfficientDet-D2 model – where small instances of microbial colonies were omitted (not localized at all), which may be caused by resizing the patches to fit the input layer size. Very low contrast between the agar substrate and the colony (*bright* subset of AGAR dataset) is an additional problem here.

Table 2. Symmetric Mean Absolute Percentage Error (sMAPE) and Mean Absolute Error (MAE) obtained for different models.

Model	Backbone	MAE	sMAPE
Faster R-CNN	ResNet-50	4.75	5.32%
Cascade R-CNN	ResNet-50	4.67	5.15%
Libra R-CNN	ResNet-50	4.21	5.19%
CBNetv2	Double ResNet-50	4.49	5.23%
YOLOv4	CSPDarknet53	4.18	5.17%
EfficientDet-D2	EfficientNet-B2	5.66	10.81%
Deformable DETR	ResNet-50	4.82	5.30%
Faster R-CNN	XCiT-T12	3.53	6.30%

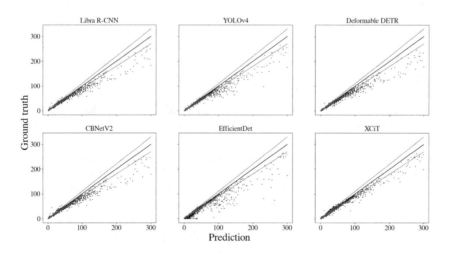

Fig. 2. The performance of microbial colony counting for 6 different models.

Table 3. Symmetric Mean Absolute Percentage Error (sMAPE) obtained for different types of microbes.

Model	E.coli	C.albicans	P.aeruginosa	S.aureus	B.subtilis
Faster R-CNN ResNet-50	5.40%	2.98%	5.53%	2.78%	1.96%
Cascade R-CNN	4.40%	2.52%	5.04%	2.51%	1.36%
Libra R-CNN	4.15%	8.49%	14.57%	3.42%	1.66%
CBNetv2	6.06%	6.03%	8.74%	2.80%	2.57%
YOLOv4	5.13%	1.99%	6.43%	2.40%	1.12%
EfficientDet-D2	3.27%	4.53%	5.59%	3.49%	1.21%
Deformable DETR	2.95%	2.56%	9.02%	2.33%	1.44%
Faster R-CNN XCiT	5.33%	5.68%	9.12%	2.24%	1.87%

5   Conclusions

In the conducted studies, we analyzed eight SoTA deep architectures in terms of model type, size, average inference time, and the accuracy of detecting and counting microbial colonies from images of Petri dishes. A detailed comparison was performed on AGAR dataset [10].

The presented results do not differ much between the different types of architectures. It is worth noting that we chose rather smaller, typical backbones for the purposes of this comparison to create a baseline benchmark for different types of detectors. It appeared that the most accurate (mAP = 0.529) and the fastest model (17 ms) is one-stage YOLOv4 network, making this model an excellent choice for industrial applications. Two-stage architectures of different types and kinds achieved moderate performance, while transformer-based architectures gave the worst results. EfficientDet-D2 turned out to be the smallest model in terms of the number of parameters.

Our experiments yet again confirm the great ability of DL-based approaches to detect microbial colonies grown in Petri dishes from RGB images. The biggest challenge here is the need to collect large amounts of balanced data. To train detectors in a fully-supervised manner, data must be properly labelled. However, identification of abnormal colonies grown in a Petri dish can be difficult even for a trained specialist - overlooked microbes were usually small and located at the edge of a Petri dish. Usually 10% error is an acceptable range for manual labeling. It is worth to point out that AI-assisted models can make predictions even for overcrowded dishes, which is a big advantage of this type of technology. They are not prone to fatigue like a human, however, variable lighting conditions can make detection even more difficult, which can be observed in our case for EfficientDet-D2 prediction for unrepresented *bright* samples.

Acknowledgements. Project "Development of a new method for detection and identifying bacterial colonies using artificial neural networks and machine learning algorithms" is co-financed from European Union funds under the European Regional Development Funds as part of the Smart Growth Operational Program. Project implemented as part of the National Centre for Research and Development: Fast Track (grant no. POIR.01.01.01-00-0040/18).

References

1. Bochkovskiy, A., et al.: YOLOv4: optimal speed and accuracy of object detection. arXiv preprint arXiv:2004.10934 (2020)
2. Cai, Z., et al.: Cascade R-CNN: delving into high quality object detection. In: 2018 IEEE/CVF Conference on Computer Vision and Pattern Recognition, pp. 6154–6162 (2018). https://doi.org/10.1109/CVPR.2018.00644
3. Carion, N., et al.: End-to-end object detection with transformers. arXiv preprint arXiv:2005.12872 (2020)
4. Chen, K., et al.: MMDetection: open mmlab detection toolbox and benchmark. arXiv preprint arXiv:1906.07155 (2019)
5. El-Nouby, A., et al.: XCiT: cross-covariance image transformers. arXiv preprint arXiv:2106.09681 (2021)
6. Girshick, R., et al.: Rich feature hierarchies for accurate object detection and semantic segmentation. arXiv preprint arXiv:1311.2524 (2014)
7. Hendrycks, D., et al.: Gaussian error linear units (GELUs). arXiv preprint arXiv:1606.08415 (2020)
8. Liang, T., et al.: Cbnetv2: a composite backbone network architecture for object detection. arXiv preprint arXiv:2107.00420 (2021)
9. Loshchilov, I., et al.: Decoupled weight decay regularization. arXiv preprint arXiv:1711.05101 (2019)
10. Majchrowska, S., et al.: Agar a microbial colony dataset for deep learning detection. arXiv preprint arXiv:2108.01234 (2021)
11. Pang, J., et al.: Libra R-CNN: towards balanced learning for object detection. arXiv preprint arXiv:1904.02701 (2019)
12. Redmon, J., et al.: You only look once: unified, real-time object detection. In: Proceedings of the IEEE Conference on Computer Vision and Pattern Recognition, pp. 779–788 (2016)
13. Redmon, J., et al.: Yolo9000: better, faster, stronger. In: Proceedings of the IEEE Conference on Computer Vision and Pattern Recognition, pp. 7263–7271 (2017)
14. Redmon, J., et al.: Yolov3: an incremental improvement. arXiv preprint arXiv:1804.02767 (2018)
15. Ren, S., et al.: Faster R-CNN: towards real-time object detection with region proposal networks. In: Proceedings of the 28th International Conference on Neural Information Processing Systems - Volume 1, NIPS 2015, pp. 91–99. MIT Press, Cambridge (2015)
16. Tan, M., et al.: Efficientdet: scalable and efficient object detection. In: 2020 IEEE/CVF Conference on Computer Vision and Pattern Recognition (CVPR), pp. 10778–10787 (2020). https://doi.org/10.1109/CVPR42600.2020.01079
17. Tan, M., et al.: EfficientNet: rethinking model scaling for convolutional neural networks. In: Chaudhuri, K., et al. (eds.) Proceedings of the 36th International Conference on Machine Learning. Proceedings of Machine Learning Research, vol. 97, pp. 6105–6114. PMLR (2019)

18. Wightman, R.: Efficientdet (pytorch) (2020). https://github.com/rwightman/efficientdet-pytorch
19. Zhu, X., et al.: Deformable DETR: deformable transformers for end-to-end object detection. arXiv preprint arXiv:2010.04159 (2020)

Low Voltage Warning System
for Stand-Alone Metering Station
Using AI on the Edge

Krzysztof Turchan$^{(\boxtimes)}$ and Krzysztof Piotrowski

IHP - Leibniz-Institut für innovative Mikroelektronik,
Im Technologiepark 25, 15236 Frankfurt (Oder), Germany
{turchan,piotrowski}@ihp-microelectronics.com
https://www.ihp-microelectronics.com

Abstract. Artificial intelligence is used in many different aspects of life these days. It is also increasingly used to create intelligent embedded devices. Their main task is to demonstrate intelligent features using limited hardware resources. This paper introduces the possibility of using AI on the Edge as a system to warn of low voltage on the battery in a stand-alone metering station powered by solar panels. The paper presents theoretical knowledge about artificial intelligence on the Edge, time series and algorithms in time series forecasting. In addition, a practical approach is presented in the Approach section. It also describes the stand-alone measurement station that was used, what type of data was collected and used, and what type of the time series forecasting algorithm was used. The main advantage of the approach is the practical use of Holt's Linear Trend method in battery voltage forecasting. Work on the project is still in progress therefore the results presented in the paper are generated during simulation.

Keywords: Artificial intelligence on the edge · Time series forecasting · Intelligent battery warning system

1 Introduction

1.1 Artificial Intelligence

It is not easy to clearly define what is artificial intelligence (AI). AI could be defined as a set of algorithms which operation is similar to intelligent human behaviour [1]. All current AI-based systems belong to the so-called Narrow intelligence [2]. Narrow intelligence is one of three defined types of AI (Narrow AI, General AI, Super AI). This type of AI is designed to perform a single task, which can not be automatically applied to other tasks in other applications. Narrow AI-based systems can be more or less complex, depending on the computing power of the device they run on. There are two approaches to implementing AI. The first approach concerns the implementation of increasingly complex artificial intelligence systems to

© The Author(s) 2022
C. Biele et al. (Eds.): MIDI 2021, LNNS 440, pp. 107–114, 2022.
https://doi.org/10.1007/978-3-031-11432-8_10

achieve the so-called General intelligence. The second approach is to implement AI algorithms on embedded devices, which are highly specialized devices designed for one or very few specific purposes and are usually embedded or included as part of a larger system. This type of AI is called AI on the Edge [3].

AI on the Edge is mainly related to above mentioned embedded devices. The idea behind this solution is to use the AI algorithms directly on the embedded device, without having to send large amounts of data to a server [3]. This solution offers many possibilities such as creating smart sensors, intelligent cameras, etc., by using various artificial intelligence methods depending on the purpose.

Forecasting is one of the most popular artificial intelligence methods whose main task is to predict future values based on data collected in the past [4]. Nowadays, predictions are made in every field where historical data is available. This is due to the fact that it gives the possibility to estimate different values, which often brings different kinds of benefits. Examples of using artificial intelligence for data prediction purposes include:

- Stock price prediction
- House price forecasting
- Weather forecasting

All of these examples have in common the fact that predictions are based on historical time-stamped data. This data is the domain of time series.

1.2   Time Series Forecasting

Time series forecasting is the process of analysing time series data using modelling and statistics to make predictions. The accuracy of the prediction depends mainly on the available data. However, an important role is played by the proper selection of the algorithm for this data. There are many algorithms for predicting time series and they are selected according to several criteria.

- **Amount of data at hand** - the more observations, the better problem understanding [4,5].
- **Time horizons and amount of variables** - the shortest time horizon, the less chance for unpredictable variables and errors [4,5].
- **Data quality** - the data must be completed, cleared and without duplication [4,5].

The most important thing is to collect data with the same interval. This allows finding trends, seasonality or cyclic behaviour. Data gaps can be filled by algorithms but this makes the final model not as accurate as it could be [5].

1.3   Forecasting Algorithms

It is not always possible that one forecasting algorithm will work equally well for different types of plots (historical data). Very often it is necessary to select an algorithm based on observations of the data. This section introduces several time series forecasting methods, along with their purpose [6].

Naive Method is a technique that assumes that the next expected point is equal to the previously observed point, due to the not very large variability of the data over time. This method is not dedicated to highly variable data sets

Simple Average is a technique that fits with data that changes over a small period of time, while the average over a longer period remains constant [6].

Moving Average. Using a simple moving average model, we predict the next value (or values) in a time series based on the average of a fixed finite number of previous values [6].

Holt's Linear Trend Method. Any dataset that follows a trend can use this method for forecasting. Thanks to the method it is possible to predict future value using a linear trend. [7]. Holt's method has three separate equations that work together to generate a final forecast. The first Eq. (1) is a smoothing equation. It adjusts the last smoothed value to the trend of the last period. The trend is updated over time using Eq. (2). It involves determining the difference between the last two smoothed values.

$$L_t = \alpha Y_t + (1 - \alpha)(L_{t-1} + T_{t-1}) \tag{1}$$

$$T_t = \beta(L_t - L_{t-1}) + (1 - \beta)T_{t-1} \tag{2}$$

$$Y_{t+1} = L_t + (h)T_t \tag{3}$$

where:
Y_{t+1} is value at time t+1,
α and β are smoothing parameters,
h is forecast horizon.

2   Motivation

Applying the techniques and methods presented in the previous section, the idea was to implement a low battery warning system for a stand-alone measurement station.

As is well known, such measurement stations are often powered by solar panels because they could be located in places far from conventional power sources. This solution could generate problems because the panels that are connected to the measurement station could not be able to keep up with charging the batteries. Of course, it is possible to install panels with a much higher power than expected, but this generates additional costs.

Instead of that a low battery warning system has been designed. The exact purpose of this system is to predict the voltage of the battery a few days ahead and to warn when the voltage level may fall below the required one. Thanks to this the user responsible for the maintenance of the station is able to react to a warning at the appropriate time.

The second problem of stand-alone measurement stations is the possibility of losing access to the network. In this case, data that could not be sent must be stored in the embedded device's memory, which is often very limited. For this reason, it is important to send as little data as possible. The most important data that must be sent is information about the continuous operation of the station and warnings about possible battery discharge. Using the forecasting algorithm on the microcontroller makes it unnecessary to send all the electrical measurement data.

3   Approach

This section provides an approach to solving the problem presented in the previous section. Furthermore, it provides information on what type of data is collected on the microcontroller and an algorithm that is responsible for forecasting. It also gives ways to optimize algorithms to save as much microcontroller processing power as possible.

3.1   Measurement Station

The warning system covered in this paper is part of a larger measurement station as shown in Fig. 1 that collects data from the environment. The whole station is powered by a battery that is charged by solar panels. The station power supply section uses a Three-Channel (3-Ch) DC meter that records voltage and current measurements data for three independent channels. 3-Ch meter is a special purpose board designed at Institute for High Performance IHP, based on MSP430F6779 microcontroller.

The first channel that is shown in Fig. 1 measures voltage and current generated on the solar panel. This allows inferring the daily insolation level. Then, the second channel is used to measure battery voltage and current. With this data, it is possible to observe the charging and discharging cycles of the battery. The last channel serves to measure the current consumption of the environment measurement system. In addition, there is a sensor for collecting the temperature data. The schematic shows also an arduino MKR1010 WiFi board that serves as an network gate for the data.

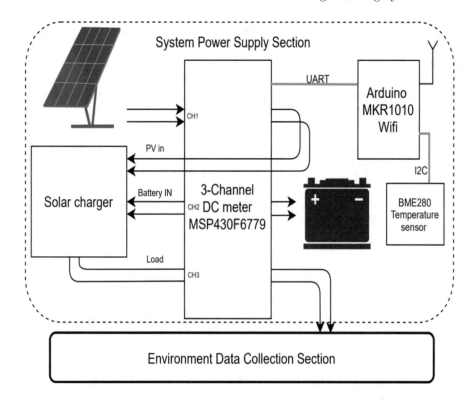

Fig. 1. Measurement station schematic

3.2   Dataset and Algorithm

This entire dataset can be used to train an advanced artificial intelligence model such as Artificial neural network (such a case will be an object of future research). However, the application must run on an embedded device with relatively low computing power and it is important to use as simple algorithm as possible. The Fig. 2 below shows an example of the data collected on the 3-Ch DC meter. As can be seen in the figure (see Fig. 2), the battery voltage series shows a trend. The Holt's method described previously is ideally suited to predict battery voltage values. This gives the possibility to generate warnings if the predicted value is below the threshold value.

The microcontroller's task is to predict the battery voltage value for the next day based on the data from the five preceding days. In order to do this, the data must be prepared in advance. Each measurement is taken every 10 min. The predicted values will be daily values so in order to standardize the data with those that will be predicted it needs to be averaged beforehand. The data is averaged every 12 h which gives 2 values per day. This data is then stored in an array with the size of 10 elements. When this array is full, the first forecast of two 12 h values is made using the Holt's Linear Trend method. The value from the second forecast value is then checked. If this value is below the battery threshold then a low battery warning is given. Each day, the two oldest elements are removed from the array to make space for the two new values. The data in the array is shifted and the two new average values, calculated from the current day are written to the array. This is followed by a recalculation of the two forecast values for the following day. Then the algorithm is repeated over and over again. This gives a daily voltage forecast for the next day.

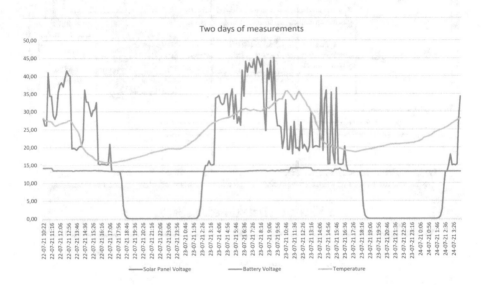

Fig. 2. Measurements collected on 3-Ch meter

The Fig. 3 shows a simulated forecast based on data that shows a downward trend. The data reflects a situation where the battery is continuously discharged by the load connected to it. It also represents the fact that battery voltage can be predicted using Holt's Linear Trend Method. The first version of the application was implemented in java programming language and executed on a PC. This was to verify the correctness of the prediction before implementing the algorithm on an embedded device. The plot shows a situation where the predicted voltage is close to 10 V. This would result in a low battery voltage warning. The prediction was made using data from the measuring station that was equipped with an under-sized solar panel on a cloudy day.

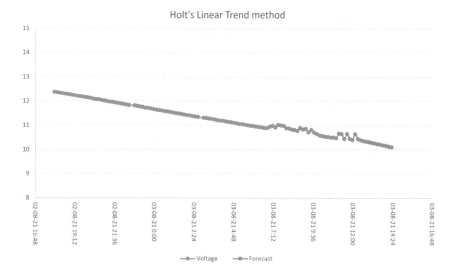

Fig. 3. Low voltage warning simulation

4   Conclusions

The whole idea of designing a low battery warning system based on artificial intelligence was to optimize the number of solar panels used in a stand-alone measuring station. Calculating the power demand and selecting the number of solar panels is nothing special, but here comes the aspect of random events such as a few days of bad weather. A warning system provides the user with sufficient time to react and secures the operation of the measuring station. As the implementation process is still underway, there are no calculated mean square errors of the forecast yet. Checking the correctness of the solution is the next task that will be performed. Unfortunately, it takes time to perform the test under natural conditions. Work will be continued directly on the microcontroller, which will give more accurate data on the speed of the proposed algorithm, needed RAM and processing power. After this, the next step will be to implement an algorithm that will use all the data described in the dataset and algorithm section. This will give the possibility to determine the prediction quality of the described solution.

Artificial intelligence is not just about complex systems for object recognition, it's also about simple algorithms implemented on devices that serve simple everyday tasks like collecting data from sensors.

Acknowledgments. This work was supported by the European Regional Development Fund within the BB-PL INTERREG V A 2014-2020 Programme, "reducing barriers - using the common strengths", by project SmartRiver, grant number 85029892 and by project SpaceRegion, grant number 85038043 and by the German government under grant 01LC1903M (AMMOD project). The funding institutions had no role in the design of the study, the collection, analyses, or interpretation of data, the writing of the manuscript, or the decision to publish the results.

References

1. Hurbans, R.: Artificial Intelligence Algorithms, 1st edn. Manning Publications Co., New York (2021)
2. Naudé, W., Dimitri, N.: The race for an artificial general intelligence: implications for public policy. AI Soc. **35**(2), 367–379 (2019). https://doi.org/10.1007/s00146-019-00887-x
3. Huh, J., Seo, Y.: Understanding edge computing: engineering evolution with artificial intelligence. IEEE Access **7**, 164229–164245 (2019). https://doi.org/10.1109/ACCESS.2019.2945338. https://ieeexplore.ieee.org/abstract/document/8861030
4. Chatfield, C.: Time-Series Forecasting, 1st edn. Chapman & Hall/CRC, Bath (2000)
5. 11 Classical Time Series Forecasting Methods in Python. https://machinelearningmastery.com/time-series-forecasting-methods-in-python-cheat-sheet/. Accessed 10 Sept 2021
6. Armstrong, J.S.: Selecting forecasting methods. In: Armstrong, J.S. (ed.) Principles of Forecasting. International Series in Operations Research and Management Science, vol. 30. Springer, Boston (2001). https://doi.org/10.1007/978-0-306-47630-3_16
7. Forecasting with a Time Series Model using Python: Part Two. https://www.bounteous.com/insights/2020/09/15/forecasting-time-series-model-using-python-part-two/. Accessed 10 Sept 2021

Test Framework for Collaborative Robot Graphene-Based Electronic Skin

Jan Klimaszewski[1](✉) ⓘ, Łukasz Gruszka[1,2] ⓘ, and Jakub Możaryn[1] ⓘ

[1] Department of Mechatronics, Institute of Automatic Control and Robotics, Warsaw University of Technology, ul. Sw. A. Boboli 8, 02-525 Warsaw, Poland
{jan.klimaszewski,lukasz.gruszka.dokt,jakub.mozaryn}@pw.edu.pl
[2] Easy Robots, Sp. z o.o., ul. Gdyńska 32/14, Radom, Poland

Abstract. Collaborative robots are one of the key pillars of Industry 4.0. Thanks to improved sensors, they can cooperate with people in a common workspace safely. Equipping the robot with an electronic skin allows increasing its safety level. The article presents the hardware and software framework of the newly developed graphene-based electronic skin for a collaborative robot. Functional laboratory tests confirm the effectiveness of the e-skin integration with the robot control system.

Keywords: Collaborative robotics · Electronic skin · Graphene nanoplatelets · Force-sensitive resistors · Robot operating system

1 Introduction

The last decades brought new human-robot collaboration (HRC) areas in the same working space with examples in the industry, education, agriculture, healthcare services, security, and space exploration [1]. The collaborative robots (so-called cobots) are an important part of the 4th industrial revolution, better described as Industry 4.0 [7]. Unlike the well-known industrial robots, cobots are simple in operation, lightweight, reliable with enhanced sensors and easy to deploy. They are designed to follow special safety measures, handling human collision detection and avoidance contrary to closed industrial robotic cells.

Most cobots adopt compliant mechanisms and lightweight designs to reduce the impact force once collisions occur. Other solutions are power and force limiting (PFL) methods based on temporal dynamic models, and current signals from embedded torque sensors [8]. Compared with the above methods, equipping cobots with the electronic skin (e-skin) is a complementary way to increase their safety levels [4,5].

The sense of touch in robotic devices has been developed for many years. Mostly as a part of a gripper or an end-effector mechanism. Adding touch allows bio-mimicry to robots, as it has a very important defence function in nature.

C. Biele et al. (Eds.): MIDI 2021, LNNS 440, pp. 115–121, 2022.
https://doi.org/10.1007/978-3-031-11432-8_11

In animals, when the harmful factor causing the pain is triggered, the body automatically moves to avoid contact with the pain-causing factor. This movement is unconditional (it occurs automatically and is not subject to will). Recently, with the sensor miniaturization and new tactile technologies development, there is a possibility for the full robot body tactile e-skin manufacturing, and usage [4].

The paper aims to describe the developed hardware and software framework, combined in a laboratory test stand, allowing the functional tests of the newly developed graphene-based e-skin solution with the cobot. During tests, the e-skin was separated physically from the robot.

The paper is organized as follows. In Sect. 2, the prototype consisting of cobot, e-skin and software integration is presented. In Sect. 3, the preliminary tests are described and the obtained results are given. Finally, Sect. 4 provides the summary and further investigation proposal.

2   Developed Test Framework

The hardware framework of the developed laboratory test stand consisted of the e-skin based on graphene nanoplatelets and an ES5 robot from EasyRobots company[1]. For tests, the e-skin module was placed outside the working space of the manipulator. It was connected to the robot via a desktop computer with an Ethernet interface. Data exchange between the computer and the e-skin was based on UART communication using the USB port. The software framework was developed using Robot Operating System (ROS[2]) [2,6].

2.1   E-skin

The e-skin was constructed of force-sensitive resistors (FSR) connected in a rectangular matrix registering a pressure map from the touch exerted on it.

The e-skin consisted of two layers, presented in Fig. 1 [4,5]. The first one is a conductive layer of comb electrodes (1b-1) printed on plastic foil and connected along columns and rows. The second one consisted of FSR sensors arranged in a rectangular pattern (1b-2) placed on a plastic foil. In the research, we have used FSR matrices with the size of a single sensor approx. 5×5 mm. In order to present the touch measurement, a matrix of FSR sensors with the dimensions of 16×32 cells was used.

The FSR matrix combined the individual FSR sensors into columns and rows. In order to estimate the location of the pressure applied to the FSR matrix, each of the columns and rows was separately, sequentially powered by a dedicated electronic controller supplying $V_{cc} = 3.3$ V to only one sensor. The output U_x (see Fig. 1a) of the matrix was connected to the ground via a measuring resistor. Accordingly, a circuit consisting of only one column and one row could be closed by the selected FSR sensor pressed to a row and column comb electrodes.

---

[1] https://easyrobots.pl/.
[2] https://www.ros.org/.

Then it was possible to read the measuring signal U_x from the selected sensor. It has been estimated that as a result of the pressure, the measured FSR resistance changed in the range of $300\,\Omega$–$3.5\,k\Omega$. Each FSR was based on graphene nanoplatelets similar to that described in [3]. The FSR matrix was connected with comb electrodes to detect the place and the contact force exerted thereon.

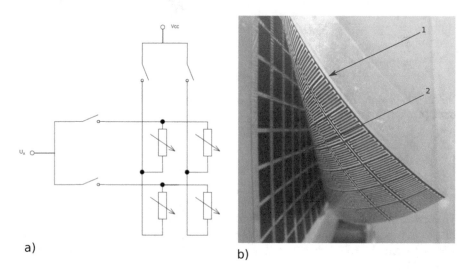

a)   b)

Fig. 1. a) Measurement principle for FSR matrix of size 2×2 [4]; b) the e-skin [5]: 1 - a conductive layer of comb electrodes printed on plastic foil, 2 - FSR sensors arranged in a rectangular pattern placed on a plastic foil.

2.2 Collaborative Robot

We have tested the e-skin prototype using an ES5 robot. It is an industrial 6-axis robot with a ROS-based communication interface, mainly used in the metalworking industry, for palletizing, packaging and welding (Fig. 2c).

2.3 Software Integration

In the research, we have used a data exchange protocol based on publishing and subscribing messages provided by ROS. As part of the software framework, the ROS node was developed, allowing the operation of the e-skin with the robot. The data package was cyclically sent (with a frequency 50 Hz) in the form of a two-dimensional matrix of numbers corresponding to the e-skin prototype surface fields, with values reflecting the pressure level. A separate ROS node associated with the robot control system was analyzing the received values seeking maximum values. If it exceeded a fixed threshold value, the node sent the information into a robot control system to slow down the robot movements.

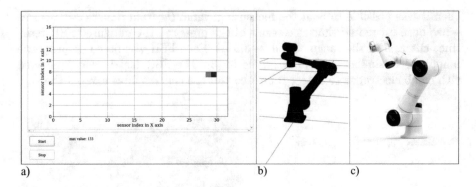

Fig. 2. The e-skin software (a), robot in simulation mode (b), a real ES5 industrial robot (c).

3   Tests

3.1   Research Method

We carried out initial research using a robot simulation mode and a prototype of the e-skin module. We observed the interaction of the developed e-skin prototype (Fig. 2a) with the robot system using visualization in RViz (ROS Visualization, see Fig. 2b).

The second stage of the research was the integration of the device with the real robot. We have tested the correctness of the algorithm for forcing the robot to switch to slow speed mode in the effect of human contact with the surface of the e-skin and finding the robot's reaction time to a change in the state of the signal transmitted from the e-skin module[3].

In both research stages, the robot moved along an experimental trajectory consisting of sections of straight lines run through the tool centre point (TCP) with a maximum speed of 1 m/s. Then, in response to human contact with the e-skin prototype matrix, the robot was forced to slow down to 250 mm/s (TCP speed, following the guidelines in point 5.7.3 of the PN-EN ISO 10218-1 standard, defining the collaborative robot reduced speed).

3.2   Research Results

Figure 3 shows the result of the robot reaction time to the pressure from the touch exerted on the e-skin module. It indicates the moments of pressure detection, receiving this information by the robot, and achieving the reduced speed. The individual times are shown in the Table 1. It can be seen that the total time from the detection of the force applied to the e-skin until the robot reaches the reduced speed is 0.414 s.

---

[3] Movie from the experiment is published on the page: http://www.jakubmozaryn.esy.es/?page_id=2418.

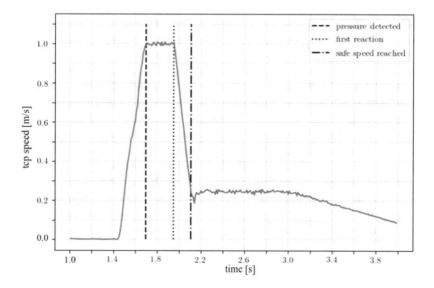

Fig. 3. Reaction times for pressure detection

Table 1. Time reactions for e-skin pressure detection.

Event	Time
Pressure detected	1.696 s
Signal received	1.950 s
Robot safe speed reached	2.110 s

4 Summary

The paper describes the hardware and software framework for the newly developed graphene-based e-skin connected with an ES5 robot manipulator. The e-skin was connected to the robot main computer unit within the Robot Operating System. The functional tests on a laboratory stand showed that the robot could be forced to slow down to 250 mm/s for the TCP speed in response to human contact with the e-skin. The presented tests prove that the developed system can be used for further research of the developed e-skin technology, as it also allows for new ways of interacting with electrical devices of daily use and can improve safety in traffic communication when introduced as an additional body coating.

The developed test framework will be the basis for further integration of the robot control system with the e-skin, which will allow for safe human-machine cooperation. A more elaborate robot response based on human touch is possible but requires improvements in the presented system's performance. In further works, the analysis and optimization of measurement and computational delays should be considered. The robot's reaction time is currently too low for human to work with it comfortably. In terms of the robot's response to human touch, the reaction time should be shortened to millisecond values.

Another part of further research will also cover the integration of the e-skin module with the robot control system by covering the whole surface of the robot structure with the flexible e-skin shell. For this purpose, a flexible cover with the applied sensory system, the same as the prototype described in this work, will be printed on the elastic mat, and a unique wiring system will be developed.

Finally, in presented research, we do not model the current flowing through sensors. It was observed that if many e-skin sensors are pressed simultaneously, the largest current will flow through the selected sensor, and a smaller current will also flow through the other sensors. It can lead to the ambiguity of the readouts. However, tests demonstrating the operation of the e-skin module have shown that it is still possible to determine the location and estimate the value of the pressure applied to the selected FSR. Addressing this problem is planned as part of the further e-skin module development.

References

1. Bloss, R.: Collaborative robots are rapidly providing major improvements in productivity, safety, programing ease, portability and cost while addressing many new applications. Ind. Robot Int. J. **43**(5), 463–468 (2016). https://doi.org/10.1108/IR-05-2016-0148
2. Ding, C., Wu, J., Xiong, Z., Liu, C.: A reconfigurable pick-place system under robot operating system. In: Chen, Z., Mendes, A., Yan, Y., Chen, S. (eds.) ICIRA 2018. LNCS (LNAI), vol. 10985, pp. 437–448. Springer, Cham (2018). https://doi.org/10.1007/978-3-319-97589-4_37
3. Janczak, D., Słoma, M., Wróblewski, G., Młożniak, A., Jakubowska, M.: Screen-printed resistive pressure sensors containing graphene nanoplatelets and carbon nanotubes. Sensors **14**(9), 17304–17312 (2014). https://doi.org/10.3390/s140917304
4. Klimaszewski, J., Janczak, D., Piorun, P.: Tactile robotic skin with pressure direction detection. Sensors **19**(21), 4697 (2019). https://doi.org/10.3390/s19214697. https://www.mdpi.com/1424-8220/19/21/4697
5. Klimaszewski, J., Władziński, M.: Human body parts proximity measurement using distributed tactile robotic skin. Sensors **21**(6), 2138 (2021). https://doi.org/10.3390/s21062138. https://www.mdpi.com/1424-8220/21/6/2138
6. Maruyama, Y., Kato, S., Azumi, T.: Exploring the performance of ROS2. In: Proceedings of the 13th International Conference on Embedded Software, pp. 1–10 (2016)
7. Sherwani, F., Asad, M.M., Ibrahim, B.: Collaborative robots and industrial revolution 4.0 (IR 4.0). In: 2020 International Conference on Emerging Trends in Smart Technologies (ICETST), pp. 1–5. IEEE (2020). https://doi.org/10.1109/ICETST49965.2020.9080724
8. Svarny, P., Tesar, M., Behrens, J.K., Hoffmann, M.: Safe physical HRI: toward a unified treatment of speed and separation monitoring together with power and force limiting. In: 2019 IEEE/RSJ International Conference on Intelligent Robots and Systems (IROS), pp. 7580–7587. IEEE (2019)

Realization of a Real-Time Decision Support System to Reduce the Risk of Diseases Caused by Posture Disorders Among Computer Users

Enes Gumuskaynak[1], Faruk Toptas[1], Recep Aslantas[1], Fatih Balki[1], and Salih Sarp[2(✉)]

[1] Aurora Bilisim, Istanbul, Turkey
{enes.gumuskaynak,faruk.toptas,recep.aslantas,
fatih.balki}@aurorabilisim.com
[2] Virginia Commonwealth University, Richmond, USA
sarps@vcu.edu

Abstract. Nowadays, diseases caused by posture disorders are becoming more common. This situation reduces the working efficiency of people, especially computer users. This study aims to provide prevention of diseases caused by posture disorders faced by computer users and realize an application software to reduce disease risks. With this realized application, computer users' movements are monitored through the camera, and the situations that may pose a risk of disease for the users are determined. Realized application software is a decision support system. This decision support system provides users suggestions to change their position according to their instant postures and supports them to work more efficiently. The user data is collected by processing the images taken from a camera using the developed computer vision algorithm. Two-dimensional (2D) human exposure estimation is performed with the obtained data. The situations that can decrease the working efficiency are specified with the data obtained from exposure estimation using the developed model. As a result of these findings, increasing the working efficiency is provided by informing the user in real-time about the situation that may decrease the working efficiency.

Keywords: Computer vision with deep learning · Computer vision with pose estimation · Decision support system

1 Introduction

At the point brought by technology, the computer has become one of the indispensable parts of our lives. The time spent in front of the computer increased with the change of both remote working and education [1]. Depending on this situation, diseases such as tendinitis, myofascial pain syndrome, neck hernia, pulmonary embolism, eye strain, and humpback caused by postural disorders continue to increase and decrease the quality of life and work efficiency [2]. Recent advances in Artificial Intelligence (AI) provide new approaches in many industries [3, 4]. In this study, pose estimation is made by processing

© The Author(s) 2022
C. Biele et al. (Eds.): MIDI 2021, LNNS 440, pp. 122–130, 2022.
https://doi.org/10.1007/978-3-031-11432-8_12

the images taken from the computer's camera to prevent the mentioned diseases. The pose estimation was developed with reference to the OpenPose [5] study, which was developed using deep neural networks and shared as open-source in python language. This study uses a bottom-up approach. Although this approach works slower than the top-down approach, it has been preferred. After all, it gives more accurate results because it can capture spatial dependencies between different people that require global inference. Posture analysis is performed using values obtained from the pose estimation study by the authors in [6]. The decision support system works by sending notifications to the user which also appears in the notification center in the case of situations that pose a risk.

In order to increase the efficiency of the pose estimation process, it is sufficient to determine only the eyes, ears, nose, shoulders, and neck to be used in the study. For this reason, a new data set containing the necessary body parts from the MPII Human Pose [7] dataset is curated. This study's customized pose estimation network is obtained by finetuning the model in [5] with the dataset, using the transfer learning method. Real-time posture analysis is performed using the obtained model. As a result of the analysis, situations that may decrease working efficiency are notified to the user in real-time to increase the user's well-being. In order for the study to be easily used by everyone, a user interface has been designed so that the user can customize it in every aspect. Improvements are made by making the designed interface used by people of different ages and professions.

2   Related Works

Object detection algorithms improved significantly with the help of AI [8]. Pose estimation is the detection of body parts in images or videos. Most studies [9–16] use a top-down approach for detection and estimation. As the first step in this approach, the person's part is detected, and then the body parts are estimated. Finally, pose estimation is made for each person. Although this approach applies to single-person detection, it cannot capture spatial dependencies between different people that require global inference. Studies [2] and [17] use a bottom-up approach in pose estimation. In this approach, all parts of each individual in the image are identified. Then, parts of different individuals are combined and grouped. Although this approach works slower than the top-down approach, it gives more accurate results because it can capture spatial dependencies between different people that require global inference [2]. Figure 1 presents an example of these approaches. For this reason, the bottom-up approach was preferred in this study.

Fig. 1. Top: typical top-down approach. Down: typical bottom-up approach [26, 27].

3 Methodology

A real-time decision support system application is developed for the well-being of computer users. In the OpenPose [5] study, which has proven its success for the purpose, the pose estimation model was retrained using the transfer learning technique utilizing the customized MPII Human Pose [18] dataset.

3.1 Environment

The camera used for image transfer is the bisoncam nb pro, which is one of the models that comes internally in personal computers. The reason for choosing this model is to show that the system can work without any extra financial cost. This is a standard camera with a fixed focus, field of view of 60°, and maximum resolution of 720 p/30 fps. Nvidia Tesla T4 graphics card specially designed for AI applications was used to shorten the time for the model training phase. The high-end graphical processor has Turing architecture, 16 GB GDDR6 memory capacity, 320 Turing cores, and 2560 Cuda cores. For the testing phase of the pose estimation model, a system with similar characteristics to the average computer user was chosen. This system has Nvidia GTX 970 m graphics card, Intel I7-6700HQ 2.60 GHz central processor, and 16 GB DDR4L short-term memory.

3.2 Dataset

The pose detection model utilized the "MPII Human Pose" dataset via transfer learning. This dataset contains approximately 25000 images containing more than 40000 people with annotated body joints. Images were systematically collected using an established taxonomy of human activities each day. Overall, the dataset covers 410 human activities, and each image is provided with an activity tag [18]. In this study, only human exposure estimation was made without distinguishing activity.

3.3 Model Development

As with many bottom-up approaches, the proposed model first detects each person's parts (key points) in the image, then assigns the parts to different people. As in Fig. 2, the network is first used to extract features from the image using one of the models in studies [19–22] with its first few layers. A set of feature maps named F is created. Then F is fed into a network of two parallel convolutional layers. Stage-1 predicts 18 confidence maps, each representing a specific part of the human pose skeleton. At this stage, the network generates several Part Affinity Fields (PAFs) (1) L1 = φ1(F), where φ1 refers to the CNNs for inference in Stage-1. The predictions from the previous stage and the original image features F are combined and used to produce refined predictions at each subsequent stage.

Stage-2 predicts a value of 38 PAFs, which represents the degree of relationship between the parts. The successive stages are used to correct the forecasts made by each branch. Two-sided graphs were created between pairs using part confidence maps. By using PAFs values, weak links in bilateral charts are trimmed, and pose skeletons are predicted.

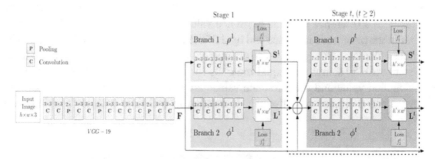

Fig. 2. Flowchart of OpenPose architecture [5].

$$L^t = \phi^\tau\left(F, L^{t-1}\right), \quad \forall 2 \le \tau \le T_P \tag{1}$$

In Eq. (1), ϕ^τ represent the CNNs for inference at stage t, and Tp is the total number of PAF stages. After Tp iterations, the process is repeated to detect confidence maps, starting with the most recent PAF estimate.

$$S^{TP} = \rho^\tau\left(F, L^{TP}\right), \quad \forall \tau = T_P \tag{2}$$

$$S^\tau = \rho^\tau\left(F, L^{TP}, S^{\tau-1}\right), \quad \forall T_P < \tau <= T_P + T_C \tag{3}$$

In Eq. (3), ρ^τ represents the number of CNNs for inference at stage t, and Tc is the number of total confidence map stages.

Confidence map results are estimated above the most recent processed PAF estimates, making a barely noticeable difference in confidence map stages. A loss function is applied at the end of each stage to direct the network to iteratively estimate the PAF values of the body parts in the first stage and the confidence maps in the second stage. L2 loss is used between the expected forecasts and the baseline information maps and fields.

$$f_L^{t_i} = \sum_{c=1}^{C} \sum_p W(p). \left\| L_c^{t_i}(p) - L_c(p) \right\|_{2'}^2 \tag{4}$$

$$f_S^{tk} = \sum_{j=1}^{J} \sum_p W(p). \left\| S_c^{tk}(p) - S_j(p) \right\|_2^2, \tag{5}$$

Equation (5) has spatially weighted the loss functions to address a practical problem where some datasets do not thoroughly label all people. Where Eq. (4) Lc is the PAF actual reference value (ground truth), Sj is the true confidence map, and is a binary mask with $W(p) = 0$ when p is missing annotation on the pixel. The mask is used during training to avoid penalizing true positive guesses. Interim inspection at each stage fills the gradient periodically, eliminating the vanishing gradient problem.

To increase the efficiency of the study, it was foreseen that it would be sufficient to determine only the eyes, ears, nose, shoulders, and neck ending to be used in the study. In order to detect only these body parts, a new data set consisting of the necessary body parts in the data set was created. With this dataset, the model was retrained using the transfer learning method to benefit the pre-trained weights of the MobilNetV2 [22].

For conditions that may pose a risk of disease, the angle values between the limbs were taken as a reference [6]. The reference values can be changed by ±30% with the slide bar in the program's interface and by ±250% in the advanced settings menu. When the values calculated due to exposure estimation went out of the reference value limits, the system took 50 samples at equal intervals for 10 s, and the average was calculated. The calculated value and the reference values are compared. When there is a situation that may pose a disease risk, the user is notified in real-time with notifications as in Fig. 3.

Fig. 3. Notification example of work in Windows 10 environment

3.3.1   Creation of the User Interface

The target audience of this study is computer users, and the user interface has been designed so that everyone can use it easily. In addition, the work has been made saved as a file with a.exe extension. The designed interface is designed in such a way that the user can customize it in every aspect, as seen in Fig. 4. The interface's main features are the selection of the posture tests, customization of wrong posture reference values, and customization of notification frequency. The settings also have five different profiles to save the made customizations.

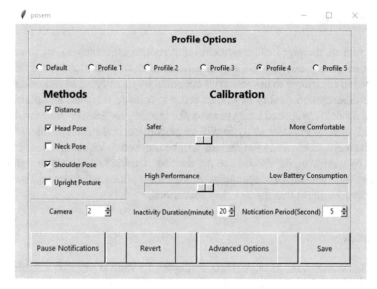

Fig. 4. Improved user interface

4   Results and Discussions

Using the images reserved from the dataset for testing, the high performed models, i.e., PersonLab [23], METU [24], Associative Emb. [25], and OpenPose [5] models were compared with the proposed method in Table 1 with respect to accuracy and precision. Since the most critical parameter for the applicability of the study is speed, the proposed method performed better than other methods.

Table 1. Performance comparison of the proposed method and the methods in the literature.

Models	Performance metrics		
	Accuracy	Precision	Speed (fps)
PersonLab	0.687	0.877	42.8
METU	0.705	0.890	53.6
Associative Emb.	0.655	0.868	31.4
OpenPose	0.642	0.862	68.2
Developed method	0.679	0.871	75.1

5 Conclusion

Health is one of the significant elements that reduce efficiency in organizations. In addition, one of the most critical expense items in most countries is healthcare. This study aimed to contribute to the country's economy by identifying situations that reduce working efficiency and quality of life. In order to achieve this aim, an optimized pose estimation model is proposed using the transfer learning technique. With the developed application, posture disorder analysis will be performed, and health problems that may occur in the waist, neck, and joint regions will be prevented. Visual disturbances will be prevented by analyzing the distance to the monitor, working environment lighting, and usage time. With the help of the proposed model, the work environment will be analyzed dynamically to support healthy working conditions.

References

1. Sarp, S., Demirhan, H., Akca, A., Balki, F., Ceylan, S.: Work in progress: activating computational thinking by engineering and coding activities through distance education. In: 2021 ASEE Virtual Annual Conference Content Access, July 2021
2. Szeto, G.P., Straker, L., Raine, S.: A field comparison of neck and shoulder postures in symptomatic and asymptomatic office workers. Appl. Ergon. **33**(1), 75–84 (2002)
3. Sarp, S., Kuzlu, M., Cali, U., Elma, O., Guler, O.: An interpretable solar photovoltaic power generation forecasting approach using an explainable artificial intelligence tool. In: 2021 IEEE Power & Energy Society Innovative Smart Grid Technologies Conference (ISGT), pp. 1–5. IEEE, February 2021
4. Sarp, S., Kuzlu, M., Wilson, E., Cali, U., Guler, O.: The enlightening role of explainable artificial intelligence in chronic wound classification. Electronics **10**(12), 1406 (2021)
5. Cao, Z., Simon, T., Wei, S.-E., Sheikh, Y.: Real-time multi-person 2D pose estimation using part affinity fields. In: The IEEE Conference on Computer Vision and Pattern Recognition (CVPR), pp. 7291–7299 (2017)
6. Wahlström, J.: Ergonomics, musculoskeletal disorders and computer work. Occup. Med. **55**(3), 168–176 (2005)
7. Hidalgo, G., Cao, Z., Simon, T., Wei, S.-E., Joo, H., Sheikh, Y.: OpenPose library. https://github.com/CMU-Perceptual-Computing-Lab/openpose
8. Sarp, S., Kuzlu, M., Zhao, Y., Cetin, M., Guler, O.: A comparison of deep learning algorithms on image data for detecting floodwater on roadways. Comput. Sci. Inf. Syst. **19**(1), 397–414 (2022)
9. He, K., Gkioxari, G., Dollar, P., Girshick, R.: Mask R-CNN. In: The IEEE International Conference on Computer Vision (ICCV), pp. 2961–2969 (2017)
10. Fang, H.-S., Xie, S., Tai, Y.-W., Lu, C.: RMPE: regional multi-person pose estimation. In: The IEEE International Conference on Computer Vision (ICCV), pp. 2334–2343 (2017)
11. Pishchulin, L., Jain, A., Andriluka, M., Thormählen, T., Schiele, B.: Articulated people detection and pose estimation: reshaping the future. In: CVPR (2012)
12. Gkioxari, G., Hariharan, B., Girshick, R., Malik, J.: Using k-poselets for detecting people and localizing their keypoints. In: CVPR, pp. 3582–3589 (2014)
13. Papandreou, G., et al.: Towards accurate multi-person pose estimation in the wild. In: The IEEE Conference on Computer Vision and Pattern Recognition (CVPR), pp. 4903–4911 (2017)

14. Chen, Y., Wang, Z., Peng, Y., Zhang, Z., Yu, G., Sun, J.: Cascaded pyramid network for multi-person pose estimation. In: The IEEE Conference on Computer Vision and Pattern Recognition (CVPR), pp. 7103–7112 (2018)
15. Xiao, B., Wu, H., Wei, Y.: Simple baselines for human pose estimation and tracking. In: Ferrari, V., Hebert, M., Sminchisescu, C., Weiss, Y. (eds.) ECCV 2018. LNCS, vol. 11210, pp. 472–487. Springer, Cham (2018). https://doi.org/10.1007/978-3-030-01231-1_29
16. Sarp, S., Kuzlu, M., Cetin, M., Sazara, C., Guler, O.: Detecting floodwater on roadways from image data using Mask-R-CNN. In: 2020 International Conference on INnovations in Intelligent SysTems and Applications (INISTA), pp. 1–6. IEEE, August 2020
17. Pishchulin, L., et al.: DeepCut: joint subset partition and labeling for multi person pose estimation. In: The IEEE Conference on Computer Vision and Pattern Recognition (CVPR), pp. 4929–4937 (2016)
18. Andriluka, M., Pishchulin, L., Gehler, P., Schiele, B.: 2D human pose estimation: new benchmark and state of the art analysis. In: The IEEE Conference on Computer Vision and Pattern Recognition (CVPR), pp. 3686–3693 (2014)
19. Simonyan, K., Zisserman, A.: Very deep convolutional networks for large-scale image recognition. CoRR, abs/1409.1556 (2014)
20. Szegedy, C., et al.: Going deeper with convolutions. In: The IEEE Conference on Computer Vision and Pattern Recognition (CVPR), pp. 1–9 (2015)
21. He, K., Zhang, X., Ren, S., Sun, J.: Deep residual learning for image recognition. In: The IEEE Conference on Computer Vision and Pattern Recognition (CVPR), pp. 770–778 (2016)
22. Sandler, M., Howard, A., Zhu, M., Zhmoginov, A., Chen, L.-C.: MobileNetV2: inverted residuals and linear bottlenecks. In: The IEEE Conference on Computer Vision and Pattern Recognition (CVPR), pp. 4510–4520 (2018). (mobilnet)
23. Papandreou, G., Zhu, T., Chen, L.-C., Gidaris, S., Tompson, J., Murphy, K.: PersonLab: person pose estimation and instance segmentation with a bottom-up, part-based, geometric embedding model. In: Ferrari, V., Hebert, M., Sminchisescu, C., Weiss, Y. (eds.) Computer Vision – ECCV 2018. LNCS, vol. 11218, pp. 282–299. Springer, Cham (2018). https://doi.org/10.1007/978-3-030-01264-9_17
24. Kocabas, M., Karagoz, S., Akbas, E.: MultiPoseNet: fast multi-person pose estimation using pose residual network. In: Ferrari, V., Hebert, M., Sminchisescu, C., Weiss, Y. (eds.) ECCV 2018. LNCS, vol. 11215, pp. 437–453. Springer, Cham (2018). https://doi.org/10.1007/978-3-030-01252-6_26
25. Newell, A., Huang, Z., Deng, J.: Associative embedding: end-to-end learning for joint detection and grouping. In: NIPS (2017)
26. Raj, B., Osin, Y.: An overview of human pose estimation with deep learning (2019). https://www.kdnuggets.com/2019/06/human-pose-estimation-deep-learning.html. Accessed 1 June 2021
27. Bogdanov, Y.: Understanding couple walking on snow near trees during daytime, 1 January 2018. https://unsplash.com/photos/XuN44TajBGo?utm_source=unsplash&utm_medium=referral&utm_content=creditCopyText. Accessed 1 June 2021

On Two Approaches to Clustering of Cross-Sections of Rotationally Symmetric Objects

Paulina Baczyńska[1]([✉]), Agnieszka Kaliszewska[1], and Monika Syga[2]

[1] System Research Institute, Polish Academy of Sciences,
ul. Newelska 6, 01-447 Warszawa, Poland
{pczubaj,agnieszka.kaliszewska}@ibspan.waw.pl
[2] Warsaw University of Technology, Faculty of Mathematics and Information Science,
Koszykowa 75, 00-662 Warsaw, Poland
m.syga@mini.pw.edu.pl

Abstract. We analyze two approaches to clustering 2D shapes representing cross-sections of rotationally symmetrical objects. These approaches are based on two ways of shape representation - contours and silhouettes - and a number of similarity measures which are based on a combination of Procrustes analysis (PA) and Dynamic Time Warping (DTW) as well as on binary matrix analysis. The comparison of efficiency of the proposed approaches is performed on datasets of archaeological ceramic vessels.

Keywords: Clustering · Shape analysis · Procrustes analysis · Dynamic time warping · Similarity measures · Contours · Binary matrix comparison · Hamming distance · Vari distance · Rogers-Tanimoto distance · Archaeological ceramic vessels

1 Introduction and Problem Formulation

Clustering methods, which rely on grouping data sets into clusters without imposing the number of clusters in advance [16,20], are widely used in many applications from natural speech recognition to biology and medicine (e.g. [13]). In order to obtain significant results of clustering, aside from the choice of the specific clustering algorithm, the following issues are of fundamental importance:

1. the choice of suitable data representation,
2. the definition of adequate similarity measure.

The problem of suitable data representation has been recognized in the literature, see e.g. [6]. The choice of data representation for the investigated data set has a considerable impact on the possibility to achieve satisfying results. Studies on the impact of shape representation on clustering architectural plans have been conducted in e.g. [17].

C. Biele et al. (Eds.): MIDI 2021, LNNS 440, pp. 131–141, 2022.
https://doi.org/10.1007/978-3-031-11432-8_13

1.1   The Problem

Our contribution is devoted to clustering 2D objects with respect to shape and size. We consider two types of shape representations: contours and silhouettes. For contour representations we use similarity measures based on a combination of DTW [1,4,15] and PA [5,9,10], as discussed in [12]. For silhouettes representations we use a number of binary distances, e.g. Hamming, Vari, Rogers-Tanimoto [18]) as well as Pattern-difference, Size-difference distance [3].

In the case of contour representation, the overall similarity measures are defined as combinations of size and shape similarity measures as proposed in [12], see Sect. 3, i.e. the clustering is performed simultaneously with respect to size and shape, whereas in the case of silhouette shape representation, we start with the clustering with respect to size, and, within the obtained (size) clusters, the clustering is next performed with respect to shape (Sect. 2).

1.2   Motivation and Contribution

The aim of the present contribution is twofold. First, by using binary matrix representation of the investigated 2D shapes we propose a clustering algorithm based on a number of discrete similarity measures combined with Affinity Propagation algorithm. Secondly, we investigate the impact of the choice of shape representation and similarity measures on the results of clustering of 2D cross-sections of rotationally symmetric 3D objects, by comparing the binary matrix approach with the contour-based approach proposed in [12].

Both approaches, the silhouette-based approach proposed in Sect. 2 and the contour-based approach proposed in [12] and recalled briefly in Sect. 3 can be applied to clustering of general 2D objects. Our motivation comes from digital cultural heritage and is related to clustering of archaeological ceramic vessels with respect to size and shape. Due to technological principles, ceramic vessels are rotationally symmetric, and so the clustering can be limited to their 2D cross sections (see Fig. 1). The papers devoted to this topic are scarce in the literature, e.g. [8,11,21].

The experiments are performed on real-life archaeological data representing cross-sections of ceramic vessels. Our entry data are in the form of black and white images of a unified resolution. When clustering archaeological ceramic data one can hardly specify the number of clusters in advance, so we are bound to use algorithms which do not require the number of clusters to be predefined.

The representations and the similarity measures, are discussed in Sect. 2 and Sect. 3. The results of experiments for archaeological ceramic data sets and the discussion are presented in Sect. 4.

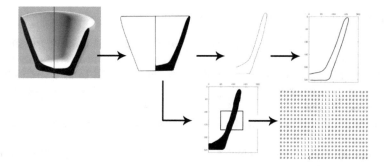

Fig. 1. The transformation of a section of rotationally symmetric object into input data.

2 Silhouette-based Approach

2.1 Data Representation

In this approach we represent shapes as matrices of pixels, to be clustered as binary matrices or, in some cases, as binary vectors. Process of preparing input data consists of two steps.

The first step relies on the analysis of the size of original input data. Size analysis is performed by KMeans cluster algorithm. To find the optimal division of the dataset, this algorithm finds clusters consisting of similar-sized objects. In the case of our experimental data, the typical outcome of KMeans algorithm, is one or two subsets-clusters of objects of similar size. This step is performed by an application written in Python 3.6. in order to retain the information about size of the investigated objects, as in the second step we perform normalization, which leads to the loss of size information. This information is recalled in the last step, before the final clustering.

Second step is related to data transformation. We start with transforming every object to the chosen fixed size, which is related to resolution of provided objects. Next step is to create matrix representation of each object. The rows and columns of the matrix represent pixels of objects. Knowing that every image consists of black and white pixels only, the matrix of pixels is created as follows: element (i, j) is given as 1, when the particular pixel is black and 0 when the particular pixel is white. This matrix represent the objects. This step is performed by application written in Python 3.6.

After data preparation, we perform clustering, based on similarity measures dedicated to binary data [3]. In this process we recall the size information (division obtained through KMeans), to be included in the final clustering result. Depending on the number of clusters obtained by KMeans (clustering with respect to size), next step, i.e. clustering with respect to shape is performed independently, i.e. in each cluster separately.

2.2 Similarity Measure

A lot of measures have been proposed for computing the similarity/dissimilarity between two vectors. In the present investigation we are interested only in those dedicated to binary input data. We base our choice on 76 binary similarity and dissimilarity measures recalled in [3]. Our selection of similarity measures for further analysis was focused on the diversity. As a result, five measures were selected for further analysis. Let X, Y - binary matrices in the same sizes. Let $x_{ij} \in X$, $y_{ij} \in Y$ be elements located at the same positions in rows and columns of matrices X and Y, respectively. Let d_{xy} be the number of instances where specific combinations of zero and ones occur: d_{00}, d_{11}, d_{01}, d_{10}. Initially, every d_{xy} value is set to 0. Next, we update them, i.e. for every pair of elements x_{ij} and y_{ij} the following operations are performed:

(i) $x_{ij} = 0 \wedge y_{ij} = 0 \Rightarrow d_{00} = d_{00} + 1$
(ii) $x_{ij} = 1 \wedge y_{ij} = 1 \Rightarrow d_{11} = d_{11} + 1$
(iii) $x_{ij} = 0 \wedge y_{ij} = 1 \Rightarrow d_{01} = d_{01} + 1$
(iv) $x_{ij} = 1 \wedge y_{ij} = 0 \Rightarrow d_{10} = d_{10} + 1$

In our clustering analysis we use the following similarity measures/distances:

1. Hamming distance: $M_H(X,Y) = d_{01} + d_{10}$,
2. Vari distance: $M_V(X,Y) = \frac{d_{01}+d_{10}}{4(d_{11}+d_{01}+d_{10}+d_{00})}$,
3. Rogers - Tanimoto dissimilarity coefficient: $M_{RT}(X,Y) = \frac{2(d_{01}+d_{10})}{d_{11}+2(d_{01}+d_{10})+d_{00}}$,
4. Pattern - Difference distance: $M_{PD}(X,Y) = \frac{4d_{01}d_{10}}{(d_{11}+d_{01}+d_{10}+d_{00})^2}$,
5. Size - Difference distance: $M_{SD}(X,Y) = \frac{(d_{01}+d_{10})^2}{(d_{11}+d_{01}+d_{10}+d_{00})^2}$,

Several characteristic features of the above similarity measures are observed:

(a) Hamming distance focuses only on information about the difference between objects. It can be successfully applied to Android malware detection ([19]).
(b) Vari's distance and Roger-Tanimoto factors analyse the full spectrum of information on object similarities and differences. It was applied to nature cases e.g. maize genetic coefficients based on microsatellite markers or diversity in olives (see e.g. [2,18,22] and the references therein).
(c) The Pattern-difference and Size-difference distance measures the squares of information;

After calculating the similarity matrix, the effectiveness of each of the similarity measures was evaluated by the clustering method. The clustering is performed by the Affinity Propagation algorithm (see e.g. [7]), which does not require a predefined number of clusters. This is an essential feature of our archaeological application. This part of clustering process is performed by application written in Python 3.6.

Below there are main algorithms used in analysis. Symbols used in algorithms:

1. $I_n[x_{I_n}, y_{I_n}]$ - original image I_n with sizes: x_{I_n} - width, y_{I_n} - height

2. $R_n[x_{R_n}, y_{R_n}]$ - resized image R_n with sizes: x_{R_n} - width, y_{R_n} - height
3. $B_n[x_{B_n}, y_{B_n}]$ - binary matrix B_n with size: x_{B_n} - number of columns, y_{B_n} - number of rows
4. N - number of images
5. D - size difference threshold
6. $d_{00}, d_{11}, d_{01}, d_{10}$ - number of instances where specific condition is passed as followed (described above)

Algorithm 1. *Single image preparation*

1: *Load image $I_n[x_{I_n}, y_{I_n}]$ to local variable*
2: *Resize image I_n to $R_n[x_{R_n}, y_{R_n}]$. After this step, all images are in the same size.*
3: *Convert image $R_n[x_{R_n}, y_{R_n}]$ to binary matrix $B_n[x_{B_n}, y_{B_n}]$ where $x_{R_n} = x_{B_n}$ and $y_{R_n} = x_{B_n}$:*
4: **for** $i = 0; i < x_{R_n}; i++$ **do**
5:    **for** $j = 0; j < y_{R_n}; j++$ **do**
6:       **if** $R_n(i,j)$ *is black* **then**
7:          $B_n(i,j) = 1$
8:       **else**
9:          $B_n(i,j) = 0$
10:       **end if**
11:    **end for**
12: **end for**

Algorithm 2. *Size clustering*

1: *Start size analysis:*
2: **for** $i = 0; i < N; i++$ **do**
3:    **for** $j = 0; j < N; j++$ **do**
4:       **if** $(x_{Ai} \cdot y_{Ai}) \div (x_{Aj} \cdot y_{Aj}) > D$ **then**
5:          *do KMeans algorithm with given number of clusters*
6:       **end if**
7:    **end for**
8: **end for**

Algorithm 3. *getInstances(B_a, B_b)*

1: *Set variables: $d_{00} = 0$, $d_{11} = 0$, $d_{01} = 0$, $d_{10} = 0$*
2: **for** $i = 0; i < x_{B_a}; i++$ **do**
3:    **for** $j = 0; j < y_{B_a}; j++$ **do**
4:       **if** $B_a(i,j) = 0 \wedge B_b(i,j) = 0$ **then**
5:          $d_{00} = d_{00} + 1$
6:       **else if** $B_a(i,j) = 1 \wedge B_b(i,j) = 1$ **then**
7:          $d_{11} = d_{11} + 1$
8:       **else if** $B_a(i,j) = 0 \wedge B_b(i,j) = 1$ **then**
9:          $d_{01} = d_{01} + 1$
10:       **else if** $B_a(i,j) = 1 \wedge B_b(i,j) = 0$ **then**
11:          $d_{10} = d_{10} + 1$

12:      **end if**
13:    **end for**
14: **end for**
15: **return** d_{00}, d_{11}, d_{01}, d_{10}

Algorithm 4. *Similarity measures*
 1: **for** $i = 0; i < N; i{+}{+}$ **do**
 2:   **for** $j = 0; j < N; j{+}{+}$ **do**
 3:     $getInstances(B_i, B_j)$
 4:     *Hamming distance:* $M_{H_{i,j}}(B_i, B_j)$
 5:     *Vari distance:* $M_{V_{i,j}}(B_i, B_j)$
 6:     *Rogers - Tanimoto dissimilarity coefficient:* $M_{RT_{i,j}}(B_i, B_j)$
 7:     *Pattern - Difference distance:* $M_{PD_{i,j}}(B_i, B_j)$
 8:     *Size - Difference distance:* $M_{SD_{i,j}}(B_i, B_j)$
 9:   **end for**
10: **end for**

Algorithm 5. *Affinity propagation*
 1: *do AffinityPropagation for set* $(M_{H_{0,1}}, M_{H_{1,0}}, ..., M_{H_{N-1,N}})$
 2: *do AffinityPropagation for set* $(M_{V_{0,1}}, M_{V_{1,0}}, ..., M_{V_{N-1,N}})$
 3: *do AffinityPropagation for set* $(M_{RT_{0,1}}, M_{RT_{1,0}}, ..., M_{RT_{N-1,N}})$
 4: *do AffinityPropagation for set* $(M_{PD_{0,1}}, M_{PD_{1,0}}, ..., M_{PD_{N-1,N}})$
 5: *do AffinityPropagation for set* $(M_{SD_{0,1}}, M_{SD_{1,0}}, ..., M_{SD_{N-1,N}})$

An important feature of the approach proposed in Algorithm 1-Algorithm 5 is that the clustering analysis is performed sequentially: first with respect to size (K-Means algorithm), and next with respect to shape (Affinity Propagation algorithm).

3   Contour-based Approach

Below we recall the contour based approach, as proposed in [12]. Contours (boundary discrete curves) are extracted from cross-sections by standard contour extraction techniques and smoothed by Savitzky-Golay filtering. Hence, objects to be clustered, are discrete curves α, i.e. pairs of vectors,

$$\alpha := (x(i), y(i)), \ i = 1, ..., k_\alpha. \tag{1}$$

3.1   Composite DTW-PA Similarity Measures

By $n \geq 2$ we denote the cardinality of the set of contours to be clustered. By i and j, $i, j = 1, ..., n$, we denote the i-th and the j-th contour, respectively. We calculate „distances" between contours i and j, $i \neq j$ by using the formulas (2), (3), (4) which measure the degree of similarity between i and j,

$$\text{Procrustes measure: } pa(i, j) := \max\{PA(i, j), PA(j, i)\}, \tag{2}$$

where $PA(i,j)$ denotes the Procrustes measure between i and j obtained from Matlab Statistics and Machine Learning Toolbox (for more details see [12]),

$$\text{direct composition measure: } dc(i,j) := \max\{DC(i,j), DC(j,i)\}, \qquad (3)$$

where $DC(i,j) = DTW(i,Z_j)$, DTW measure is calculated in Matlab Signal Processing Toolbox, and Z_j is the optimal curve obtained from $PA(i,j)$, the measure DC was introduced and analysed in [12],

$$\text{scale component measure: } \gamma(i,j) := 1 - \min\{\gamma^*(i,j), \gamma^*(j,i)\}, \qquad (4)$$

where $\gamma^*(i,j)$ is the optimal scaling factor obtained from the $PA(i,j)$ (see [12] for more details). We put $pa(i,i) = 0$, $dc(i,i) = 0$, $\gamma(i,i) = 0$. To ensure comparability of the above defined measures, we use the following normalization formulas for dc and γ

$$ndc(j) := \frac{1}{max_{i=1,...,n} dc(i,j)}, \qquad \text{for} \quad j = 1,...,n.$$

$$n\gamma(j) := \frac{1}{max_{i=1,...,n} \gamma(i,j)}, \qquad \text{for} \quad j = 1,...,n.$$

In absence of normalization we put $ndc(j) = ndc = 1$, for all $j = 1,...,n$ or $n\gamma(j) = n\gamma = 1$, for all $j = 1,...,n$.

3.2 Similarity Matrix

Now we recall the similarity measure (5) and the similarity matrix (6), which where defined and investigated in [12]. Let $\mu, \lambda, \omega \in \mathbb{R}$ be given numbers (weights). The similarity measure (SM) is given as follows:

$$SM_{ij}(\mu, \lambda, \omega, ndc, n\gamma) := \qquad (5)$$

$$\mu \cdot pa(i,j) + \lambda \cdot ndc(j) \cdot dc(i,j) + \omega \cdot n\gamma(j) \cdot \gamma(i,j).$$

Depending upon the choice of parameters $\mu, \lambda, \omega \in \mathbb{R}$ we obtain basic measures $pa(i,j), dc(i,j), \gamma(i,j)$, combinations of two or three of them, with, or without, normalization. By calculating SM_{ij} according to (5) for every pair of contours i and j, $i,j = 1,...,n$, we get the similarity matrix

$$SM(\mu, \lambda, \omega, ndc, n\gamma), \qquad (6)$$

which is symmetric and has zeros on the main diagonal. Different choices of $\mu, \lambda, \omega, ndc, n\gamma$ lead to different similarity matrices and consequently, to different clustering outcomes. In particular, we obtain

1. weighted Procrustes and scale component matrix

$$WPSM = SM(\mu, 0, \omega, 1, 1), \qquad (7)$$

138   P. Baczyńska et al.

2. weighted direct composition and scale component matrix where direct composition values are normalized

$$WNDCSM = SM(0, \lambda, \omega, ndc, 1),\qquad(8)$$

3. weighted direct composition and scale component matrix, where both direct composition values and scale component values are normalized

$$WNDCNSM = SM(0, \lambda, \omega, ndc, n\gamma).\qquad(9)$$

We perform clustering on the basis of the similarity matrix (6) by the standard hierarchical algorithms (Matlab Statistics and Machine Learning Toolbox) and generate the dendrogram. The average linkage is used to measure the distance between clusters.

An important feature of the contour-based approach is that the weights appearing in the definition of the similarity measures (5) allow the simultaneous clustering with respect to size and shape.

4   Discussion of the Results of the Experiment

Below we summarize the clustering results obtained with the help of methods from Sect. 2 and Sect. 3. The experiment has been conducted on seven sets of real-life data compiled from archaeological material. One of the datasets is presented in Fig. 2. It is worth noting that there are no benchmark data sets related to clustering of archaeological ceramics.

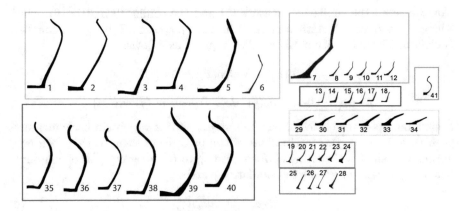

Fig. 2. Set 1

According to general basic evaluation criteria (e.g. [16]), clustering results are acceptable when the resulting clusters are well defined, i.e. the distances between elements within the cluster are small (intra clusters characteristic), while the distances between clusters are high (extra clusters characteristic). Due to the unlabeled nature of the investigated data our evaluation is expert-based.

The percentage of correctly clustered elements is summarized in Table 1. Average scores have been calculated over all sets for all methods. For readability, best scores for both classes of methods (i.e. contour- and silhouette-based) are marked in bold.

Based on the results, we feel confident to evaluate the performance of both approaches as acceptable, with the average scores for each method between 78.68%–86.63%.

Some correlations between characteristics of the data sets and the results obtained with the two approaches can be established. The situation is interesting for e.g. Sets 1. Here the contour-based approach performs well while the silhouette-based approach seems to be lacking. The opposite is true for Set 2, where the average score for contour-based approach is 85%, while the silhouette-based approach has an average score of 92%. Same is true for Set 4 with scores 86% and 94% respectively. This points to a set of characteristics of the original dataset that have a strong impact on the results obtained. Moreover, with few, set-specific exceptions, the choice of the similarity measure in silhouette-based approach, seem to have little impact on the resulting clusters. What is even more striking, also the applied orders of clustering: sequential (first size, next shape in silhouette-based approach), simultaneous (weighted size and shape in contour-based approach) seem not to influence the clustering results in a decisive way. One should note, however, that the reported computational experiment is based on sets of medium sizes (30–40 elements). Some robustness of the results with to respect sequential versus simultaneous approaches could be implied by the fact that size and shape are the features of completely different natures: the size can easily be quantified, whereas the formal quantified expression of shape is still the topic of current research https://www.dam.brown.edu/people/mumford/vision/shape.html, [14]. The characteristics impacting the results of the considered approaches remain to be investigated and are the subject of ongoing investigation.

Table 1. Percentage of correctly classified elements for Sets 1–7 according to expert evaluation. The number of elements in each set is given in brackets.

Method	Set 1 [41]	Set 2 [41]	Set 3 [45]	Set 4 [36]	Set 5 [51]	Set 6 [40]	Set 7 [40]	Average
$SM(\frac{1}{2},0,\frac{1}{2},1,1)$	82.93%	78.05%	75.56%	75%	69.23%	87.5%	87.50%	78.68%
$SM(\frac{3}{4},0,\frac{1}{4},1,1)$	85.36%	90.24%	**82.22%**	86.11%	78.43%	85%	80%	83.91%
$SM(0,\frac{1}{2},\frac{1}{2},ndc,1)$	**90.24%**	80.49%	77.78%	86.11%	78.43%	82.5%	82.50%	82.58%
$SM(0,\frac{3}{4},\frac{1}{4},ndc,1)$	87.8%	90.24%	77.78%	**91.67%**	78.43%	87.5%	**90%**	86.20%
$SM(0,\frac{1}{2},\frac{1}{2},ndc,n\gamma)$	87.8%	75.61%	75.56%	88.89%	78.43%	**92.5%**	80%	82.68%
$SM(0,\frac{3}{4},\frac{1}{4},ndc,n\gamma)$	90.24%	92.68%	80%	86.11%	82.35%	92.5%	82.50%	**86.63%**
Hamming distance	77.50%	92.68%	82.22%	94.44%	78.43%	**77.50%**	**87.50%**	84%
Vari distance	77.50%	92.68%	82.22%	94.44%	78.43%	**77.50%**	**87.50%**	84%
Rogers–Tanimoto distance	77.50%	**95.11%**	82.22%	94.44%	78.43%	**77.50%**	**87.50%**	**85%**
Size-difference distance	**80.49%**	90.24%	**87.80%**	94.44%	**80.39%**	65%	85%	83%
Pattern-difference distance	77.50%	90.24%	77.778%	94.44%	78.43%	**77.50%**	**87.50%**	83%

5 Conclusion

In conclusion, both, the contour- and the silhouette-based approaches are viable to be applied to the clustering of 2D objects with respect to size and shape. Moreover, to the best of our knowledge, the silhouette-based approach, has not yet been used in archaeological applications. In further research, the problem of computational experiments for large data sets should be addressed.

References

1. Aronov, B., Har-Peled, S., Knauer, C., Wang, Y., Wenk, C.: Fréchet distance for curves, revisited. In: Azar, Y., Erlebach, T. (eds.) ESA 2006. LNCS, vol. 4168, pp. 52–63. Springer, Heidelberg (2006). https://doi.org/10.1007/11841036_8
2. Balestre, M., Von Pinho, R., Souza, J., Lima, J.: Comparison of maize similarity and dissimilarity genetic coefficients based on microsatellite markers. Genet. Mol. Res. **7**(3), 695–705 (2008)
3. Choi, S.S.S.: Correlation analysis of binary similarity and dissimilarity measures (2008)
4. Efrat, A., Fan, Q., Venkatasubramanian, S.: Curve matching, time warping, and light fields: new algorithms for computing similarity between curves. J. Math. Imaging Vision **27**(3), 203–216 (2007)
5. Eguizabal, A., Schreier, P.J., Schmidt, J.: Procrustes registration of two-dimensional statistical shape models without correspondences. CoRR abs/1911.11431 (2019)
6. Farias, F.C., Bernarda Ludermir, T., Bastos-Filho Ecomp, C.J.A., Rosendo da Silva Oliveira, F.: Analyzing the impact of data representations in classification problems using clustering. In: 2019 International Joint Conference on Neural Networks (IJCNN), pp. 1–6 (2019). https://doi.org/10.1109/IJCNN.2019.8851856
7. Frey, B.J., Dueck, D.: Clustering by passing messages between data points. Science **315**, 2007 (2007)
8. Gilboa, A., Karasik, A., Sharon, I., Smilansky, U.: Towards computerized typology and classification of ceramics. J. Archaeol. Sci. **31**(6), 681–694 (2004)
9. Goodall, C.: Procrustes methods in the statistical analysis of shape. J. R. Statist. Soc. Ser. B (Methodol.) **53**(2), 285–339 (1991)
10. Hosni, N., Drira, H., Chaieb, F., Amor, B.B.: 3D gait recognition based on functional PCA on Kendall's shape space. In: 2018 24th International Conference on Pattern Recognition (ICPR), pp. 2130–2135 (2018)
11. Hristov, V., Agre, G.: A software system for classification of archaeological artefacts represented by 2D plans. Cybern. Inf. Technol. **13**(2), 82–96 (2013)
12. Kaliszewska, A., Syga, M.: A comprehensive study of clustering a class of 2D shapes. arXiv: 2111.06662 (2021)
13. Leski, J.M., Kotas, M.P.: Linguistically defined clustering of data. Int. J. Appl. Math. Comput. Sci. **28**(3), 545–557 (2018)
14. Mumford, D.: Mathematical theories of shape: do they model perception? In: Proceedings of Conference 1570, Society of Photo-Optical & Instrumentation Engineers (SPIE), pp. 2–10 (1991)
15. Müller, M.: Information Retrieval for Music and Motion. Springer Science, Heidelberg (2007). https://doi.org/10.1007/978-3-540-74048-3

16. Owsiński, J.W.: Data Analysis in Bi-partial Perspective: Clustering and Beyond. Studies Computational Intelligence, Springer, Cham (2020). https://doi.org/10.1007/978-3-030-13389-4
17. Rodrigues, E., Sousa-Rodrigues, D., Teixeira de Sampayo, M., Gaspar, A.R., Gomes, A., Henggeler Antunes, C.: Clustering of architectural floor plans: a comparison of shape representations. Autom. Construct. **80**, 48–65 (2017)
18. Rogers, D.J., Tanimoto, T.T.: A computer program for classifying plants. Science **132**(3434), 1115–1118 (1960)
19. Taheri, R., Ghahramani, M., Javidan, R., Shojafar, M., Pooranian, Z., Conti, M.: Similarity-based android malware detection using hamming distance of static binary features. Future Gener. Comput. Syst. **105**, 230–247 (2020)
20. Wierzchoń, S.T., Kłopotek, M.A.: Algorithms of Cluster Analysis. Institute of Computer Science, Polish Academy of Sciences (2015)
21. Yan, C., Mumford, D.: Geometric structure estimation of axially symmetric pots from small fragments. In: IASTED International Conference on Signal Processing, Pattern Recognition, and Applications, vol. 2, pp. 92–97 (2002)
22. Zaher, H., et al.: Morphological and genetic diversity in olive (olea europaea subsp. europaea l.) clones and varieties. Plant Omics J. **4**(7), 370–376 (2011)

Digital Interaction

Video Projection on Transparent Materials

Vojtěch Leischner$^{(\boxtimes)}$ and Zdenek Mikovec

Faculty of Electrical Engineering, Department of Computer Graphics and Interaction, Czech Technical University in Prague, Prague, Czech Republic
{leiscvoj,xmikovec}@fel.cvut.cz
https://dcgi.fel.cvut.cz/

Abstract. We propose a new coating for light projection on transparent materials that can open new possibilities for design. What we often struggle with is the lighting of clear glass. By definition, clear glass lets most of the light pass through. We have found a way to turn glass surfaces opaque or transparent by using ultraviolet (UV) fluorescence coating. In combination with a UVA light source, we can project the dynamic content onto a glass surface treated with a special coating that transforms the UVA light into visible light. The added benefit of such a coating is that it can be applied to any organically shaped surface using a spray gun, not just flat surfaces. Another advantage is that the light source is nearly invisible to the human eye, especially with a UV light pass-through filter. We have created a prototype with a modified overhead projector to measure the light characteristics and documented the steps to reproduce our results.

Keywords: Video projection · Fluorescence · Video mapping · Transparent materials · Glass · Transparent projection · Emissive projection display

1 Introduction

Video projection on glass or other transparent materials is a key piece of technology for augmented reality applications. By overlaying the real view with the projected image, we can add information related to the real world. Imagine projecting navigation arrows on the front glass of a car. With additional sensors, we can overlay the navigation markers on the physical road. Such a solution is inherently safer and more convenient than having to switch our focus between a traditional LCD panel and a front car window. Similar to the car window, the actual glass panel is usually not flat but curved due to aerodynamic or ergonomic demands. Even more problematic is a video projection on hand-blown glass with the curved surface in more than one direction. We have to consider the geometry of the projection surface to achieve non-deformed projections from the point of the viewer. Furthermore, we have to deal with the light distribution as we want

© The Author(s) 2022
C. Biele et al. (Eds.): MIDI 2021, LNNS 440, pp. 145–152, 2022.
https://doi.org/10.1007/978-3-031-11432-8_14

to achieve uniform brightness. Another problem that can arise with projection on transparent surfaces is reflection. We have to plan the placement of the video projector to avoid the glare affecting the viewer. Additionally, we have to reckon with reflections inside the material itself.

However, to project on a glass one needs to treat the surface so it can reflect the projected light. There are numerous solutions ranging from projection films, special materials inside the glass to special light sources. The goal is to let most of the visible light pass through but at the same time be able to project on discrete surfaces. That means we need the material to be clear for the human eye but at the same time opaque for the video projection. Most commonly available projection films sacrifice a bit of transparency and also video projection gain to achieve this. Generally speaking, more opaque the projection film is, the better gain can be achieved but worse transparency.

Our solution keeps the original transparency of the material and at the same, we can selectively emit light directly from the surface. We achieved this using a special light source - LED in UVA range combined with a special clear UV fluorescence varnish. As the projected UV light only excites the surface, the actual light that the viewer sees emanates directly from the surface where the UV is transformed into a visible light spectrum. Our projection system can be applied by spray painting, which is way more convenient to apply than projection films. It can also be easily applied in the case where projection films could not be used such as complex organic shapes. Another major benefit is maintaining the original material transparency.

2   State of the Art

We could use an active transparent LCD for video projection. Transparent LCD displays can achieve 80% light transmission [8]. However, for curved or large-scale surfaces this approach is not feasible. For such use-cases, video projection is a preferred solution.

There are numerous methods for projecting video onto clear glass. One of the solutions might be PDLC [3] also known as a smart glass film. PDLC consists of a sheet of liquid crystals that are randomly oriented and opaque. When the high voltage is applied, PDLC crystals align and turn the surface transparent. Unfortunately, commercially produced PDLC films have numerous disadvantages. It cannot accommodate organic shapes, it cannot be dimmed selectively, it is expensive and requires high voltage generators. However, it also gives good light gain.

Closer to our proposed solution are various semitransparent video projection films that are applied onto glass [10]. There are numerous technologies used, the most commonly available use reflective nanoparticles dispersed inside the film. The more particles are inside the film, the more gain it gives, but it also reduces the transparency. A disadvantage of such films is that they have to be applied on flat or curved surfaces in one direction only. It might be impractical to apply a film on a surface that is curved in more than one direction. Such films are typically cheaper than PDLC, but they suffer from worse gain compared to PDLC.

Our proposed solution is similar to the one developed by T.X. and Cheng, B [9]. In their paper, they discuss the system of three UV fluorescent films merged in a thin sheet and applied onto a glass and excited by a custom-made UV light source. However, the materials used are not described in the paper and thus it is not possible to reproduce the results. Moreover, the UV fluorescence is achieved using a film, so it is limited to flat glass applications.

Unlike T.X. and Cheng, B. solution, our system is monochrome only, but this could be extended by using different wavelength dyes with specific light sources. We also discuss DIY light source modification and use cases. Fluorescent dyes are widely available, can be mixed with different varnishes depending on your use case. Overall, the solution is cheaper compared to using films or PDLC.

3   Use Case

We intend to use the coating to achieve a dynamic light design using hand-blown glass. The final product is a chandelier consisting of multiple glass components. Such glass components are typically not flat, but they have volume and varying shapes. To project video on components, sandblasting was previously required. UV coating preserves the transparency of glass but enables video projection on the surface. Thanks to curves in more than one direction, it is not possible to use traditional projection films glued on the surface. On top of that, each component is original and scaling would also be an issue. If we would include the UV pigment inside the molten glass directly, we would lose the transparency as we have verified in a separate experiment. That is why our coating is the best solution for a given use case.

Another proposed use case is a volumetric 3D display (see Fig. 4). We have created a prototype consisting of a grid of glass tubes hanged from the ceiling. Tubes are arranged in a way to not overlap from the point of view of the video projector. Thanks to calibration using camera video projector pair [7] we matched the 3D model of the display with the video projector point of view. Therefore, we can render virtual 3D objects - when the virtual objects collide with the tubes in the display, the intersection is highlighted. By using clear glass with applied UV varnish solution, we can further improve the overall transparency of the display. See video documentation of the proposed physical 3D display prototype [6].

There are many art installations in which the semi-transparent layered surface is used for video projection. Such as a piece for Laterna Magika exhibition by Michael Bielicky [2] in Meetfactory gallery [11]. Bielicky uses flexible plastic strips hanging from the ceiling as a projection surface. Unlike the proposed 3D volumetric display, it does not work with depth or display discrete 3D images. Rather the video image is broken by the projection surfaces. Proposed UV coating enables artists to work in the third dimension while not losing transparency.

Moreover, the transparent display is also usable for augmented reality use cases. Another setup we will test the UV projection on is a glass cylinder, 3 m in diameter, where the viewer can stand inside (see Fig. 5). Using video projection, we can enable the viewer to see through the glass and display additional information on the glass surface. Such a display can be also useful for air traffic control or car HUD.

4   Projection System

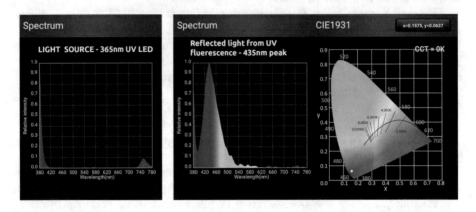

Fig. 1. Left: light source, Right: reflected light from the UV coating on clear glass.

4.1   UV Dye Varnish Applied to Glass

We have mixed Aragurad 109 UV dye made by Aralon with VA 177 TH epoxy varnish by ElChemco originally designed to protect PCB. The coating is first mixed as 45% UV dye and 55% varnish. The solution is then diluted to 10% dye and 90% varnish. We have manually mixed the solution using a wooden stick for about 30 s until the dye is uniformly distributed in the varnish. The solution can be spray painted using a regular paint gun or applied directly with a brush. The coating can also be applied selectively to paint a concrete motive on the glass (see Fig. 2) or applied to the whole surface and mask the light source or use video projection to create dynamic content. UV light 365 nm is converted into mainly blue light 445 nm. The optimal wavelength for the highest luminescence of the dye was measured at 320 nm, but we had available a 365 nm light source only for the prototype. Therefore, we anticipate that the performance of the varnish can be further improved by using an optimal light source. By using different dyes we can achieve different colors as well. More readily available fluorescence compounds can be also used. Such as quinine diluted in the appropriate varnish solution.

4.2   Safety

Powerful UVA light sources such as used LED can cause skin aging and eye damage when exposed for prolonged periods of time [1]. To remedy this, we suggest using UV blocking varnish applied on the other side of the glass to

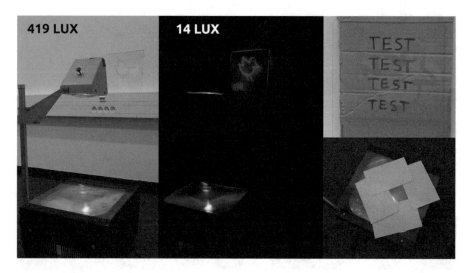

Fig. 2. Left: 3 mm clear glass with motive painted with UV coating under ambient light, Middle: same as left but in dark, Right: overhead projector with masked output. Text behind glass to demonstrate transparency.

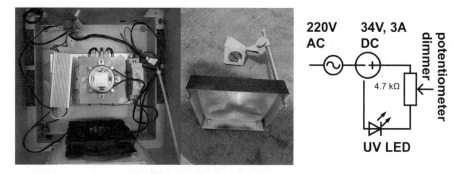

Fig. 3. Left: modified overhead projector with UV LED, Middle: overhead projector optics (Fresnel lens and adjustable mirror), Right: electrical schema.

Fig. 4. 3D glass tube display prototype with one video projector as a light source.

Fig. 5. Left: modular glass cylinder for video projection. Right: car front window video projection illustration

ensure no or minimum of UV light passing through. However, even without it much of the UV light is turned into visible light thanks to the proposed UV coating. In any case, the light source should be placed in a manner that it will not spot directly on the person. For example, it could point from above at an angle to the glass surface or be used as a back projection. Light spill outside the targeted area should be minimized by masking the light source.

4.3   UV Light Source

We are using readily available SMD LED (10 * 10 grid) UV 365 nm 100 W by OTdiode driving it at 34 V 3000 mA. We have measured the light characteristics using a spectrometer (see Fig. 1 left) *Lighting passport* by Asensetech with 8 nm optical resolution. The LED is fitted inside an overhead projector (see Fig. 3). We have removed the original light source and replaced it with our LED, a heatsink with an active cooler rated for 180 W (Sunon LM310-001A99DN), an appropriate 100 W LED driver and multirotary potentiometer for dimming. The overhead projector also includes a large Fresnel lens, an optical lens, and a mirror that direct the image to the screen. Such a system can be used to apply various masks that can be created from paper or by painting with a black marker onto a cellophane sheet. We have chosen an overhead projector for the simplicity of the light source modification. Have a look at the dynamic projection of bubbles inside water placed over the projector [5].

Other methods that can be used are modifying or creating a custom DLP projector. The LED projector is not suitable as the LED display typically also includes a UV filter that would block the light. Even if you remove a UV filter, the LED display will suffer from UV light exposure and eventually degrade. Therefore, DLP technology is much more suitable. We are currently working on modifying the DLP projector by replacing the projector lamp. Sensors detecting the open cover have to be bypassed and it is also necessary to simulate the circuit communication with the video projector motherboard to bypass the lamp manufacturer detection. You could also construct your own DLP projector to avoid these issues.

Another option is to use a moving headlight with gobo wheels. Gobos are circular stencils made from metal or glass mounted on light optics that can rotate. With such a light source you can selectively cover a 360° view and with custom gobo wheels you can project images or text. By using two gobo wheels rotating in opposite directions you can achieve effects similar to an animated fire or water surface. Such projectors are already commercially available [4]. Motorized light barn doors can be used as well.

5   Future Research

We will focus on measuring the relative gain of the fluorescence varnish we have created. We will also try to come up with a formula for a suitable UV dye. The next step is to create an actual prototype for each use case mentioned, such as augmented reality, hand-blown glass design chandelier, and a physical 3D display consisting of multiple Plexiglas sheets. We also want to switch from using an overhead projector to a custom-made DLP video projector.

6   Conclusion

We have developed a new projection system for clear glass or other transparent materials. The main benefit is that the coating can be applied in the form of spray paint and thus allow for non-flat video projection surfaces. Another advantage is that the light source is nearly invisible (especially when used with a UV pass-through filter) and the light appears to emit from the projected surface. Different UV dyes and varnishes can be used to adapt to a particular use case. The proposed solution is also cheaper compared to PDLC or projection films. We have measured the light characteristics of the coating and documented our application.

Acknowledgement. This research has been supported by the project funded by a grant SGS19/178/OHK3/3T/13 and by RCI (CZ.02.1.01/0.0/0.0/16_019/0000765).

References

1. Ahmad, S.I.: Ultraviolet Light in Human Health, Diseases and Environment, vol. 996. Springer, Cham (2017). https://doi.org/10.1007/978-3-319-56017-5
2. Bielicky, M.: Welcome: Michael Bielicky. https://www.bielicky.net/

3. Doane, J., Golemme, A., West, J.L., Whitehead, J., Jr., Wu, B.G.: Polymer dispersed liquid crystals for display application. Mol. Cryst. Liq. Cryst. **165**(1), 511–532 (1988)
4. Rosco Laboratories: Image spot®uv (2021). https://us.rosco.com/en/product/image-spot-uv
5. Leischner, V.: Dynamic water projection with UV fluorescence (2021). https://youtu.be/VPlQcuon6BM. Accessed 19 Aug 2021
6. Leischner, V.: Tube display experiment (2021). https://youtu.be/TdCM8wDTmnc. Accessed 19 Aug 2021
7. Moreno, D., Taubin, G.: Simple, accurate, and robust projector-camera calibration. In: 2012 Second International Conference on 3D Imaging, Modeling, Processing, Visualization & Transmission, pp. 464–471. IEEE (2012)
8. Okuyama, K., et al.: 79-4L: late-news paper: highly transparent LCD using new scattering-type liquid crystal with field sequential color edge light. In: SID Symposium Digest of Technical Papers, vol. 48, pp. 1166–1169. Wiley Online Library (2017)
9. Sun, T.X., Cheng, B.: A new emissive projection display technology and a high contrast DLP projection display on black screen. In: Emerging Digital Micromirror Device Based Systems and Applications III, vol. 7932, p. 793209. International Society for Optics and Photonics (2011)
10. Ye, Y., Liu, Z., Chen, T., et al.: Toward transparent projection display: recent progress in frequency-selective scattering of RGB light based on metallic nanoparticle's localized surface plasmon resonance. Opto-Electron. Adv. **2**(12), 12190020 (2019)
11. Česálková, L., Svatoňová, K.: Laterna magika: the memory of an experiment, September 2019. https://bit.ly/38IL6Ii

Balancing Usability and Security of Graphical Passwords

Kristina Lapin[(⊠)] and Manfredas Šiurkus

Vilnius University, Vilnius, Lithuania
`kristina.lapin@mif.vu.lt, manfredas.siurkus@mif.stud.vu.lt`

Abstract. Although most widely used authentication involves characters as passwords, but secure text-based passwords are complex and difficult to remember. Users want to have easy to remember passwords, but these are vulnerable to various kinds of attacks and are predictable. To address these problems, graphical passwords that involve selection of images and drawing lines have been proposed. The encoded images support creation of secure passwords and facilitate their memorability. Thus, they are considered as an alternative to strengthen password security while preserving usability because secure textual passwords become more complicated to use. The research addresses the issue of improving the user experiences during graphical authentication. This paper examines cued recall-based, pure recall-based and recognition-based approaches. The proposed scheme is based on recognition-based scheme that is selected as the least vulnerable to various attacks. The solution is currently under development, two qualitative usability testing sessions are performed and the participants' feedback is discussed.

Keywords: Graphical authentication · Recognition-based schemes · Recall-based schemes · Cued recall-based schemes

1 Introduction

A good password needs to be easy to remember and hard to guess [1]. These are contradictory requirements that encourage research on balancing security and usability of the user authentication. Textual passwords still dominate over the other methods of end-user web authentication in web applications due to their simplicity and affordability [2]. Typical authentication includes an email address and a secret alphanumeric text-based password that is cost-effective in implementation and familiar to users. However, this scheme requires significant mental effort when the user follows all security conventions for secure password creation and further usage. Saltzer and Schroeder identify the psychological acceptability as important design principle that ensures effective security [6]. There are many examples of the fact that overly complex security systems actually reduce effective security [7].

Graphical passwords are a type of knowledge-based authentication that attempt to leverage the human memory for visual information with the shared secret being related to or composed of images or sketches [3, 4]. They offer a good alternative to text-based passwords in terms of memorability and security [5]. Discussion on authentication

© The Author(s) 2022
C. Biele et al. (Eds.): MIDI 2021, LNNS 440, pp. 153–160, 2022.
https://doi.org/10.1007/978-3-031-11432-8_15

schemes produced alternative methods, but a comparison of their security and usability attributes does not show better results comparing with widely used textual password schemes [2].

We focus on the graphical recognition-based approaches as they are aimed at strengthening security and enhancing usability. The goal of this paper is to enhance recognition-based technique that will reduce their main drawback, namely the possibility of shoulder surfing attack while preserving the high security.

This paper examines known issues of existing authentication methods and propose the preliminary solution. This paper is structured as follows. Section 2 deals with the graphical authentication schemes that are classified according to required mental efforts and actions. Section 3 describes the proposed enhancement and findings of the first qualitative studies. Finally, the conclusions are drawn.

2   Related Works

The users create textual passwords using words because want easy to rememberable secrets. Such passwords are vulnerable to dictionary attacks [9]. Exploration of a person's social information reveals their relative names that makes possible social engineering attacks [10–12].

The idea behind graphical passwords is to leverage human memory for visual information, with the shared secret being related to or composed of images or sketches [4]. The recognized drawback of graphical passwords is shoulder surfing when someone captures the password while watching over the user's shoulder during the entering [1, 13]. There are many attempts to overcome this challenge.

Graphical passwords are classified according the required mental efforts or metrics based on user actions. Classification based on mental efforts comprises recognition-, recall- and cued recall-based approaches [4, 13].

Cued-recall approach involves images to create an association with words that facilitate creation of a stronger textual password. An example of this approach is InkBlot authentication [14] where the user has to associate the randomly looking images with memorable textual character or pair of them. Associations support the creation of strong textual random character password. Observation of the shown images does not help the attacker because the association outcome cannot be observed. As all textual passwords, the InkBlot secret string can be cracked using keystroke logger. From the memorability perspective, the advantage is that such a scheme facilitates memorability of complex password elements. However, the users can still forget more sophisticated password creation rules because the users may not necessarily select the first letter of the associated word and link it to security word. This may result with difficulties recalling the password elements [15]. All in all, cued-recall schemes are better than textual passwords because of their resilience to brute force and dictionary attacks and partially facilitated memorability. However, the security problem remains with keystroke logging; usability drawback can occur when the user uses a sophisticated password creation rule.

Popular examples of recall-based approaches are Draw-A-Secret (DAS) [16] and PassGo [17] authentication methods where the latter can be viewed as a discretized version of the DAS [16]. The user draws a continuous stroke or several strokes on chosen elements of the grid or set of points. According to involved user actions this approach is classified also as a draw-metric authentication scheme [18]. The strokes are graphical, therefore easy to memorize. However, a study on widely used variation of DAS – the Android pattern lock – revealed that users are inclined to draw simple pictures, such as "L"; as a result users get weak passwords [19]. Also, input pattern is susceptible to guessing [20], shoulder surfing [1, 13], smudge[21], thermal [22] and video-based [23] attacks. An interesting improvement to these challenges is proposed in behavioral pattern lock approach [24] in which the users do not need to create their own pattern and memorize them. Instead, during the login, the public patterns are shown along with guidance on how to draw them. This approach acquires touch dynamics from the touch screen and sensors, extracts useful features that classify users using machine learning.

Pure recognition-based approaches present set of pictures. The users are expected to recognize the secret pictures, selected during the registration. Popular examples are Passfaces [25] and Déjà vu [26]. Passfaces exploit the human brain's ability to quickly recognize familiar faces, Déjà vu is based on the ability to remember previously seen images [13]. During login the user points the password image in several rounds, therefore, these schemes are also classified as loci-metric or click-based graphical password schemes [18].

Generally, recognition-based schemes are resilient to most of known attacks, but the shoulder surfing [2]. Passfaces can also be predictable because the attractiveness, gender and race of faces can affect the user's choice [20]. Convex Hull Click (CHC) [1] scheme guards against shoulder-surfing attacks by human observation, video recording, or electronic capture. Similar to Passfaces, CHC requires several rounds of challenge-response authentication. In CHC the pass-icons serve as the points, and the edges are lines visualized in the user's mind. To respond to the challenge, the user clicks anywhere within the convex hull. This is a difference with Passfaces because during CHC login the user never points directly the password icons. However, the rearrangement of the login screen, finding the pass-icons, forming the mental convex in mind requires time and mental efforts. Summarizing, advantages of recognition-based schemes are based on easier recognition of the password. However, most recognition-based schemes are vulnerable to shoulder surfing, in some cases to guessing. Although CHC scheme makes unable observation and guessing, it requires significant time and mental effort.

All graphical authentication schemes facilitate the memorability in various ways. Cued recall-based approaches involve images that help to remember the complex and unguessable textual password, but finally the user enters the textual password. Therefore, the problems with keystroke logging and memorability still remain. Recall-based schemes also support memorability but they are more vulnerable to the attacks comparing to recognition-based approaches (Table 1). Recall-based schemes require to draw the same pattern, so it can be observed, recorded and used in attacks. Recognition-based schemes involve clicking on recognized images, but each time images are presented in different positions. Therefore, the potential attack can be only shoulder surfing. Guessing occurs only for faces.

Table 1. Comparison of the graphical authentication schemes.

Scheme	User actions	Security drawbacks
Recall-based	Clicking on an image points in determined order	Shoulder surfing, guessing, smudge, thermal and video-based attacks
Cued-recall	Images associations facilitate remembering the complex textual password	Keystroke logging, memorability
Recognition-based	Selection of image or face from the provided set	Shoulder surfing, guessing (in special conditions)

In conclusion, the safest graphical passwords belong to a recognition-based group. Therefore, we further examine the way to refine their security and usability.

3 Proposed Graphical Password

Our goal is to propose improvement of the recognition-based schemes, such as Passfaces that are vulnerable to guessing and shoulder surfing attack. Our idea is to combine positive aspects of recognition-based schemes to avoid guessing and to minimize opportunities to observe it. Passfaces scheme is vulnerable to guessing because of the user's preferences to choose faces of a particular race, gender and attractiveness. Déjà vu passwords involve randomly generated images. We suggest to use gallery of photos that do not contain faces and are not randomly generated. Our assumption is that photos can be more pleasurable to use and easier to recognize comparing with randomly generated images.

Known recognition-based schemes make several image recognition rounds to strengthen security. However, this prolongs the logging and increases efforts required to authenticate. When authentication is frequently needed, it can irritate the user and discourage from using a service. To minimize the number of selection rounds we suggest to present photos on the one screen and ask the user to tap on password images. After each tap the images are reordered. This protects from shoulder surfing because while tapping the image is covered with the finger; after the operation the other image is shown on the tapped place. Because images are rearranged after each selection, the same image can be tapped several times.

To enhance security, the number of shown images can be increased. By tapping on the image, the associated complex textual string that fit security requirements is entered. The encoding of photos can be easy to figure out by trying several different passwords. Therefore, to further improve the security of this solution, the use of password hashing is proposed. This prevents the data leakage from the password database at hacking.

The registration scenario involves choosing the images that should be recognized during the login. The user enters a login name and chooses at least three images from the interactive photo gallery during registration (Fig. 1). The order of selections is also fixed. To confirm the registration the user repeats the choice of images with the same order. It is allowed to enter only two identical photos from the required at least three. During the login the user enters login name and selects the images in the same order.

Fig. 1. The registration window asks two times to select pictures in the same order (on the left) and login interface (on the right)

At login, password photos are encrypted. Each photo corresponds to 5 characters, the password contains numbers, uppercase and lowercase letters, and characters not described in the alphabet. During the registration, the user is asked to select at least 3 photos that generate a 15-character password that is considered safe to brute force attacks [27, 28]. The implemented solution also uses cryptography, the password is encrypted using the Bcrypt library before an account is created.

If the user forgets a secret, the new password can be set using the user's e-mail. This solves the problem when a user forgot password or when the account is locked after too many incorrect attempts to log on to the system.

We conducted usability testing with 5 participants aged 20 to 58 years with various IT usage experience levels. The goal of qualitative study was to collect their opinions about how easy is to memorize the photos selected during the registration and to select them while logging. Initially, the users were exposed with a prototype that contained 30 photos. This number did not fit on the screen, so the screen had to be scrolled. Testers complained that it was difficult to find the photo after each tap. They were distracted and get confused if they didn't find the right photo on the initial screen and had to scroll.

On the second session the number of photos was reduced to fit the space above the page fold. We found that no more than 20 photos should be shown. With a larger gallery, you should consider the size of the photos to make sure they fit on the high definition screens. The photos should be reduced or enlarged according to the screen resolution. This redesign presented better results comparing to the initial version.

Testing participants requested to provide feedback while selecting an image. However, marking selected password photos is not secure because of possibility to observe.

Instead of marking the selected image, we suggest to provide the number of already selected photos. Moreover, the explanation of registration and logging procedure should be provided for the new users.

4 Conclusions

Comparison of the graphical authentication methods revealed that recognition-based approaches are the least vulnerable to attacks. This work aimed at providing a more efficient way of ensuring security while not scarifying usability. Our focus was to improve the recognition-based approach that is based on the brain's natural ability to recognize visual information. The developed scheme does not involve human faces, thus protecting against the threat of password prediction. Shoulder surfing chances are reduced due to rearrangements of the photo gallery after each selection. The user opinions during the testing revealed that this authentication is easily learned and requires an acceptable amount of efforts during registration and further authentication.

The usability and security of this scheme directly depend from the number of photos and selection rounds. The proposed scheme requires to choose at least 3 photos. This minimal case forms a 15-character length string that fits the security requirements.

The further usability studies are needed to conduct quantitative research that would allow to compare the efficiency of registration and logging with existing recognition-based methods.

References

1. Wiedenbeck, S., Waters, J., Sobrado, L., Birget, J.-C.: Design and evaluation of a shoulder-surfing resistant graphical password scheme. In: Proceedings of the working conference on Advanced visual interfaces, pp. 177–184. Association for Computing Machinery, New York (2006). https://doi.org/10.1145/1133265.1133303
2. Bonneau, J., Herley, C., van Oorschot, P.C., Stajano, F.: The quest to replace passwords: a framework for comparative evaluation of web authentication schemes. In: 2012 IEEE Symposium on Security and Privacy, San Francisco, CA, USA, pp. 553–567. IEEE (2012). https://doi.org/10.1109/SP.2012.44
3. Nelson, D.L., Reed, V.S., Walling, J.R.: Pictorial superiority effect. J. Exp. Psychol. Hum. Learn. Mem. 2, 523–528 (1976). https://doi.org/10.1037/0278-7393.2.5.523
4. Biddle, R., Chiasson, S., Van Oorschot, P.C.: Graphical passwords: learning from the first twelve years. ACM Comput. Surv. (CSUR). 44, 1–41 (2012)
5. Kayem, A.V.D.M.: Graphical passwords – a discussion. In: 2016 30th International Conference on Advanced Information Networking and Applications Workshops (WAINA), pp. 596–600 (2016). https://doi.org/10.1109/WAINA.2016.31
6. Saltzer, J.H., Schroeder, M.D.: The protection of information in computer systems. Proc. IEEE 63, 1278–1308 (1975). https://doi.org/10.1109/PROC.1975.9939
7. Dourish, P., Redmiles, D.: An approach to usable security based on event monitoring and visualization. In: Proceedings of the 2002 Workshop on New Security Paradigms, pp. 75–81. Association for Computing Machinery, New York (2002). https://doi.org/10.1145/844102.844116
8. Platt, D.: The Joy of UX: User Experience and Interactive Design for Developers. Addison-Wesley Professional, Boston (2016)

9. Morris, R., Thompson, K.: Password security: a case history. Commun. ACM. **22**, 594–597 (1979). https://doi.org/10.1145/359168.359172
10. Ramanan, S., Bindhu, J.S.: A survey on different graphical password authentication techniques. Int. J. Innov. Res. Comput. Commun. Eng. **2**(12), 7594–7602 (2014)
11. Krombholz, K., Hobel, H., Huber, M., Weippl, E.: Advanced social engineering attacks. J. Inf. Secur. Appl. **22**, 113–122 (2015). https://doi.org/10.1016/j.jisa.2014.09.005
12. Yıldırım, M., Mackie, I.: Encouraging users to improve password security and memorability. Int. J. Inf. Secur. **18**(6), 741–759 (2019). https://doi.org/10.1007/s10207-019-00429-y
13. Sarohi, H.K., Khan, F.U.: Graphical password authentication schemes: current status and key issues. IJCSI **10**, 437 (2013)
14. Stubblefield, A., Simon, D.: Inkblot authentication (2004)
15. Shnain, A.H., Shaheed, S.H.: The use of graphical password to improve authentication problems in e-commerce. In: AIP Conference Proceedings, vol. 2016, p. 020133 (2018). https://doi.org/10.1063/1.5055535
16. Jermyn, I., Mayer, A., Monrose, F., Reiter, M.K., Rubin, A.D.: The design and analysis of graphical passwords. In: Proceedings of the 8th USENIX Security Symposium, Washington, D.C., p. 15 (1999)
17. Tao, H.: Pass-Go, a new graphical password scheme (2006)
18. Sharma, A., Dembla, D., Shekhar: Implementation of advanced authentication system using opencv by capturing motion images. In: 2017 International Conference on Advances in Computing, Communications and Informatics (ICACCI), pp. 759–765 (2017). https://doi.org/10.1109/ICACCI.2017.8125933
19. Andriotis, P., Tryfonas, T., Oikonomou, G.: Complexity metrics and user strength perceptions of the pattern-lock graphical authentication method. In: Tryfonas, T., Askoxylakis, I. (eds.) HAS 2014. LNCS, vol. 8533, pp. 115–126. Springer, Cham (2014). https://doi.org/10.1007/978-3-319-07620-1_11
20. Davis, D., Monrose, F., Reiter, M.K.: On user choice in graphical password schemes. In: Proceedings of the 13th USENIX Security Symposium, San Diego, CA (2004)
21. Aviv, A.J., Gibson, K., Mossop, E., Blaze, M., Smith, J.M.: Smudge attacks on smartphone touch screens. In: Proceedings of USENIX Conference on Offensive Technology (WOOT), pp. 1–7 (2010)
22. Abdelrahman, Y., Khamis, M., Schneegass, S., Alt, F.: Stay cool! Understanding thermal attacks on mobile-based user authentication. In: Proceedings of the 2017 CHI Conference on Human Factors in Computing Systems, pp. 3751–3763. Association for Computing Machinery, New York (2017)
23. Ye, G., et al.: Cracking android pattern lock in five attempts. In: Proceedings of the 2017 Network and Distributed System Security Symposium 2017 (NDSS 2017). Internet Society, San Diego, California, USA (2017)
24. Ku, Y., Park, L.H., Shin, S., Kwon, T.: Draw it as shown: behavioral pattern lock for mobile user authentication. IEEE Access **7**, 69363–69378 (2019). https://doi.org/10.1109/ACCESS.2019.2918647
25. Two Factor Authentication, Graphical Passwords - Passfaces
26. Dhamija, R., Perrig, A.: Déjà Vu: A User Study Using Images for Authentication. Presented at the (2000)
27. Komanduri, S., et al.: Of passwords and people: measuring the effect of password-composition policies. In: Proceedings of the SIGCHI Conference on Human Factors in Computing Systems, pp. 2595–2604. Association for Computing Machinery, New York (2011)
28. Kelley, P.G., et al.: Guess again (and again and again): measuring password strength by simulating password-cracking algorithms. In: 2012 IEEE Symposium on Security and Privacy, pp. 523–537 (2012). https://doi.org/10.1109/SP.2012.38

Interdisciplinary Research with Older Adults in the Area of ICT: Selected Ethical Considerations and Challenges

Kinga Skorupska[1,2,3(✉)] [iD], Ewa Makowska[1,2] [iD], and Anna Jaskulska[1,3] [iD]

[1] Polish-Japanese Academy of Information Technology, Warsaw, Poland
[2] SWPS University of Social Sciences and Humanities, Warsaw, Poland
[3] Kobo Association, Warsaw, Poland
kinga.skorupska@pja.edu.pl

Abstract. In this paper we analyse, classify and discuss some ethical considerations and challenges related to pursuing exploratory and interdisciplinary research projects in the area of ICT, especially those involving older adults. First, we identify spotlight areas, which are especially prominent in these fields. Next, we explore possible pitfalls interdisciplinary researchers may stumble onto when planning, conducting and presenting exploratory research activities. Finally, some of these are selected and discussed more closely, while related open questions are posed.

Keywords: Older adults · ICT · Research practice · Research ethics

1 Introduction

Ensuring equal access and opportunity, across the board, for people to benefit from and contribute to digital services and content creation is a challenge of an ethical nature in itself. There is still much research to be done in the realm of Information and Communication Technologies (ICT) for older adults with ever more to come as new technologies emerge. Yet, there too are some ethical pitfalls anyone researching this area has to stay aware of. Even the more recent studies in Human-Computer Interaction (HCI) with older adults may be tainted by prevailing stereotypes related to health, social life and ICT skills of participants, as noticed by Vines et al. [30] and the tendency to consider designing for older adults in terms of mainly accessibility [10] resulting in problems in sustainable ICT solutions' development [8,9].

Addressing the widening digital divide [17] is important not only to allow older adults to engage with cutting-edge technologies but also to empower them to use ICT-based solutions such as online banking, e-health services, e-commerce or e-learning. This is the reason we have researched novel ways of interacting with ICT solutions, such as Smart TV-based interfaces for crowdsourcing [24,25] or ones that are chatbot-based [23]. Furthermore, in the context of Smart Home

© The Author(s) 2022
C. Biele et al. (Eds.): MIDI 2021, LNNS 440, pp. 161–170, 2022.
https://doi.org/10.1007/978-3-031-11432-8_16

Technology (SHT) we explored Voice User Interfaces (VUI) [6,16] and Brain-Computer Interfaces (BCI) [12]. We have also conducted a VR co-design study with older adults [15] and created a checklist of factors influencing the Immersive Virtual Reality (IVR) experience for participants of all ages [21] to make study replication easier. We have also participated in projects evaluating older adults' interaction with online citizen science tasks [22] and engaged older adults in participatory design activities [14], especially at the frontiers of HCI [11].

The considerations we discuss in this paper derive from our research and professional experience, both in the academic and business contexts. It is our hope that they will contribute to the discussion of the adverse effects that rigid scientific practices and insufficient interdisciplinary cooperation may have on the quality of research with older adults at the frontiers of HCI. Researching older adults' preferences regarding digital interaction in an open-minded and ethically sensitive way will help not only mitigate barriers to their use of technology-mediated solutions, but will also help consider older adults' strong suits in new ICT solutions' designs.

2   Selected Considerations

There are general ethical considerations related to good research conduct and fair reporting, for example, those outlined in European Code of Conduct for Research Integrity[1] developed by ALLEA - the European Federation of Academies of Sciences and Humanities, however, each field has its own specificity, and there are clear ethical guidelines on the design of ICT systems [26] with Value Sensitive Design (VSD) coming to mind [2], just as there are multiple ethical considerations related to research and practice involving participants in general, and older adults in particular [3,4,20].

2.1   Older Adults and Technology

However, it is at the intersection of these areas that some challenging ethical questions arise. To bring them to light we have created a SWOT-inspired overview of the selected items in our research field related to its strengths, limitations, opportunities and challenges. The overview is visible in Fig. 1. The discussion that follows is based on the interaction of these items and it includes the observations related to challenges and some dilemmas we have encountered in our work so far, as well as ones which feel as though they may become our future concerns.

---

[1] https://allea.org/code-of-conduct/.

STRENGTHS	LIMITATIONS
1. creating solutions that foster digital inclusion, serving as a gateway to improve ICT-literacy 2. making it easier to contribute remotely for a group underrepresented online 3. developing solutions with older adults in a participatory manner, drawing insights from them directly	1. older adults are a very heterogeneous demographic group 2. easier access to user groups not representative of the general population (interested in ICT) 3. overcoming stereotypes of the researchers 4. overcoming self-stereotypes of the participants

OPPORTUNITIES	CHALLENGES
1. exploring the ageing process and the opportunities coming with strong suits of older adults (e.g. increase in crystallized intelligence [18]) 2. discovering older adults' preferences regarding the interaction with novel technologies 3. designing systems that match older adults' strong suits and aspirations 4. exploring and understanding older adults' diverse needs and motivations	1. constant awareness of one's own unconscious biases 2. empowering older adults' to evaluate ICT-solutions with confidence 3. avoiding the reinforcement of the filter bubble 4. using jargon-free clear communication to ensure informed consent 5. designing with very diverse groups of older adults in mind 6. appropriately addressing the user needs that may come with old age (e.g. lower working memory [32]) 7. IT system design challenges (user privacy, clear communication, accessibility, bias-free content)

Fig. 1. Overview of key considerations regarding ICT research with older adults.

2.2   Interdisciplinary Cooperation

Although the ethical path for each researcher may seem clearly paved by high profile publications in each discipline and the best practices taught, at the intersection of multiple disciplines this image begins to crack. Interdisciplinary cooperation is challenging first because of jargon or skills which are meant to be complementary, but often serve to divide into those in-the-know or not (which is just a matter of time and willingness to learn). It is also a challenge to become aware of, bring to light and discuss different internalized research practices, goals and expectations, which may be surprising - but can be the result of formal reporting requirements, best practices considered common for each discipline and skills and individual preferences of the researchers.

Research Priorities. Such collaboration is also difficult because of the subtle differences in the order of values prioritized, which at first may be difficult to realize. For example, from the point of view of ICT-system design, ensuring that recruitment procedures screen potential users for mild cognitive decline goes against the idea of accessible design and preventing ICT-exclusion. However, it is a common practice in some other disciplines, as has its reasons and its place. But when screening procedures are applied automatically, without taking into account the type of research being done (exploratory) and what is its end goal (preventing digital exclusion) then misunderstandings may appear.

Vocabulary. Even the vocabulary may differ. For example, while everyone speaks of "participants", for some fields this term is interchangeable with "users", while in others it is common to think of "subjects" - and this actually has deep implications when it comes to the perception of the reasons for doing research and the attitude towards the participants, which may be objectifying them: if we study users, we observe them to gather insights to design something better for them, if we study subjects, then why are we studying them? It takes us one step further from them, but at what point such distance means detachment rather than objectivity?

Outdated Standards. Another difficult area is connected to standards and procedures which often feel rigid. This can be felt with standard validated questionnaires, which could use an update. This is both in terms of design and content, as language evolves, habits and expectations change, and what in the past may have indicated a problem, now may be a result of reliance on technology (e.g. not remembering one's location on the map, or the date). At a time when we are armed with best practices used commercially for UX writing, design and accessibility, it is a pity some of these updates are yet to reach the academic practice, and are guarded by the ideas of consistency with past research (should we be consistent with sub-par practices?) or the effort needed to validate new designs. Redesigning study documentation and tools could improve understanding, thus allowing them to better perform their function (by testing a closer approximation of reality, rather than the ability to fill out forms) and address some of the concerns related to informed consent (e.g. using images and diagrams which could improve understanding instead of, or along, blocks of text [27]).

Business, Public Institutions and NGOs. Businesses often view older adults through the lens of stereotypical problems related to health and maintaining a reasonable standard of independent living (AAL – Ambient and Assisted Living). This happens despite perceiving the potential of the silver economy, as the employees in organizations attempting to address the needs of older adults are often younger and unfamiliar with scientific findings in this area [13]. This situation is aggravated by diverse expectations and practices [31]. Businesses and industry expect fast iterations and immediate outcomes - reflected in the agile approaches to project management [29]. Meanwhile, the research process is similar to the waterfall project management methodology [19]. This approach is reinforced by grant applications, which often require researchers to present the whole project plan with expected outcomes and KPIs, the declaration of which, without prior research and verification may be affected by stereotypes. Such focus on KPIs is also conductive to treating project participants, including older adults, in an objectifying way, to engage them just enough to meet the project numbers requirements. A similar problem of focusing on indicators is faced by public institutions and NGOs implementing tasks and projects aimed at older adults. In such a case, there is little space left for an individualized approach to the participants.

2.3   Participants

Information Portioning. There is a trend in research, encouraged by a strict understanding of a research practice to keep the real point of an experiment hidden [28], to say as little as possible about the research being done. However, this leads to a few problems, even if we disregard the discussion of whether informed consent is possible in such cases. For example, if little information is disclosed participants may attempt to guess what the research may be about, and in this, change their behaviour in an unpredictable way. This happens often with older adults as they expect to be tested on the stereotypical problems, such as balance, hearing, eyesight or cognitive performance. It is even more prominent if task design somehow reinforces this perception. So, if a task somehow involves listening, participants may want to prove to the researcher that their hearing is fine and change the behaviour they exhibit, which may affect the actual point of the experiment.

Distance and Detachment. If researchers place themselves too far from the participants either by using objectifying language in participant-facing situations, or really any jargon, or by forgetting to check in with work and life outside of the academia, they may underuse their empathy, as it comes with some cognitive costs [1], and loosen their connection to reality. This fault is especially evident with endeavours which are meant to have positive social impact addressing problems in contexts the researchers are not very familiar with. One example here would be designing unrealistic seeming simulations for the experiments [7] (e.g. based on work problems that do not appear often, using speech examples which are rigid and sound unnatural, creating tasks to evaluate items the general population would have little interest in). Such problems could be addressed by introducing the practice of participatory design, in which members of the target group could co-design research scenarios and evaluate their realness and relevance, to make research closer to life. We believe this co-design step could be a very valuable addition to the experiment design process.

Managing Expectations. Yet again, keeping one's defined distance from participants is necessary in some types of research to provide unbiased results, however, when in participatory design the study participants become team members and co-designers, then the appropriate distance is harder to pinpoint. It involves the difficult process of navigating the intricacies of each interaction while always thinking of the good of the participants first.

Staying alert when using unscripted communication is important. Especially if the research project has no direct implications for the lives of the participants engaged with it, it is very important to communicate this clearly, as false expectations of working towards immediate personal benefit may appear, especially among vulnerable groups, such as some older adults. The same consideration was also prominent in our previous project concerning migration[2] as it is crucial to not engage the resources of people, especially those with fewer resources, in an unclear context.

2.4   Researchers

Answering Real Needs. Just like it is important to verify study design in participatory design workshops, especially when threat of stereotyping is present, there comes one more step, which should actually be taken before that. It is necessary to confront the idea of the research and expected solution with the real needs of the potential study participants, unless conducting basic or foundation research. In the business world this is done by doing marketing research, customer segmentation, interviews, ethnographic studies - which all come before the commitment to projects [5,31]. However, in research, the decision to do a project may be facilitated by available grants and quite often, relies on previous literature analysis, which may be not as applicable to the specific situation where the project will take place.

Unconscious Biases. Another challenge is realizing and facing one's own unconscious biases and entering system or experiment design with a mind free of untested or poorly-backed assumptions. One example we encountered was the expectation that older adults will not perform a longer text transcription as part of a crowdsourcing study - which was shown to be false in the study, only because in the end the task was included [22]. Actually, it is exactly the constant awareness of our haste to make use of shortcuts and prevailing narratives that can help us spot such study design faults at an early stage, thus allowing us to gain unexpected insights.

Prioritizing Research Outcomes. This brings us to the replication crisis and the danger of flawed studies influencing the design choices of future research,

---

[2] "Advanced Learning and Inclusive Environment (ALIEN) for Higher Education through greater knowledge and understanding of migration flows in Europe" was a project we conducted between 2016–2019. See more at: https://alienproject.pja.edu.pl/.

thus creating a self-reinforcing cycle of confirmation bias. Again, it is our job as researchers to evaluate the validity of previous studies. This is why there is great value in well performed meta-analyses, for example using the PRISMA approach[3]. For this reason, the constant need and pressure to do original research is also concerning, as sometimes our resources are best spent on doing a proper meta-analysis to update our understanding allowing us (and others) to design better studies and refresh our curriculum, and necessarily teaching methods, to benefit our students (who ought to be thought as our future colleagues) - which could have the biggest impact on our research area and its progress.

3   Conclusions

An important ethical question each scientist ought to answer is their role in the society and, in a broader sense, in the world. To what extent should scientists study what is applicable and practical? How far can researchers follow their passion and interests and at what point these become just an exercise of a curious mind, rather than a way to contribute to social good and positive change? Ideally, these goals would be aligned, and curiosity would lead researchers to socially valuable contributions which are immediately applicable in collaboration with policy makers, businesses and industry.

The barriers we encounter in driving research-informed progress go deep. They are related to the structure of the educational system, research world, incentive design and the broader reality we function in as these tend to be rigid. These are challenges that are often too great for any researcher to single-handedly address, so they are also one of the reasons to maintain a healthy network of collaborators—both in the academia, in our discipline, in other disciplines and out in the real world, working, aspiring and facing problems. Such people not only can help make our research more relevant, to help us meet today's complex and interdisciplinary challenges. They can also guide us through the intertwined web of ethical considerations to take into account, not only when designing and conducting research, but also taking it beyond the confines of the academia.

Acknowledgments. We would like to thank the many people and institutions gathered together by the distributed Living Lab Kobo and HASE Research Group (Human Aspects in Science and Engineering) for their support of this research. In particular, the authors would like to thank the members of XR Lab Polish-Japanese Academy of Information Technology and Emotion-Cognition Lab SWPS University as well as other HASE member institutions.

References

1. Empathy is hard work: People choose to avoid empathy because of its cognitive costs. J. Exp. Psychol. Gener. **148**(6), 962–976 (2019). https://doi.org/10.1037/xge0000595

---

[3] http://www.prisma-statement.org/.

2. Albrechtslund, A.: Ethics and technology design. Ethics and Information Technology **9**, 63–72 (2007). https://doi.org/10.1007/s10676-006-9129-8
3. Diaz-Orueta, U., Hopper, L., Konstantinidis, E.: Shaping technologies for older adults with and without dementia: reflections on ethics and preferences. Health Inform. J. **26**(4), 3215–3230 (2020). https://doi.org/10.1177/1460458219899590. (pMID: 31969045)
4. Gros, A., et al.: Recommendations for the use of ICT in elderly populations with affective disorders. Front. Aging Neurosci. **8**, 269 (2016). https://doi.org/10.3389/fnagi.2016.00269, https://www.frontiersin.org/article/10.3389/fnagi.2016.00269
5. Hewitt-Dundas, N., Gkypali, A., Roper, S.: Does learning from prior collaboration help firms to overcome the 'two-worlds' paradox in university-business collaboration? Res. Policy **48**(5), 1310–1322 (2019)
6. Jaskulska, A., et al.: Exploration of voice user interfaces for older adults–a pilot study to address progressive vision loss. In: Biele, C., Kacprzyk, J., Owsiński, J.W., Romanowski, A., Sikorski, M. (eds.) Digital Interaction and Machine Intelligence, pp. 159–168. Springer International Publishing, Cham (2021)
7. King, M.: Chapter 6 - the challenge of research. In: King, M. (ed.) Psychology in and Out of Court, pp. 82–100. Pergamon, Amsterdam (1986). https://doi.org/10.1016/B978-0-08-026798-2.50010-9, https://www.sciencedirect.com/science/article/pii/B9780080267982500109
8. Knowles, B., Hanson, V.L.: Older adults' deployment of 'distrust'. ACM Trans. Comput. Hum. Interact. **25**(4), 21:1–21:25 (2018). https://doi.org/10.1145/3196490
9. Knowles, B., Hanson, V.L.: The wisdom of older technology (non)users. Commun. ACM **61**(3), 72–77 (2018). https://doi.org/10.1145/3179995, http://doi.acm.org/10.1145/3179995
10. Knowles, B., et al.: The harm in conflating aging with accessibility. Commun. ACM **64**(7), 66–71 (2021). https://doi.org/10.1145/3431280, https://doi.org/10.1145/3431280
11. Kopeć, W., et al.: Participatory design landscape for the human-machine collaboration, interaction and automation at the frontiers of HCI (PDL 2021). In: Ardito, C., et al. (eds.) INTERACT 2021. LNCS, vol. 12936, pp. 564–569. Springer, Cham (2021). https://doi.org/10.1007/978-3-030-85607-6_78
12. Kopeć, W., et al.: Older Adults and Brain-Computer Interface: An Exploratory Study. Association for Computing Machinery, New York, NY, USA (2021). https://doi.org/10.1145/3411763.3451663
13. Kopeć, W., Nielek, R., Wierzbicki, A.: Guidelines towards better participation of older adults in software development processes using a new spiral method and participatory approach. In: Proceedings of the 11th International Workshop on Cooperative and Human Aspects of Software Engineering, pp. 49–56. CHASE 2018, Association for Computing Machinery, New York, NY, USA (2018). https://doi.org/10.1145/3195836.3195840
14. Kopeć, W., Skorupska, K., Jaskulska, A., Abramczuk, K., Nielek, R., Wierzbicki, A.: Livinglab PJAIT: towards better urban participation of seniors. In: Proceedings of the International Conference on Web Intelligence, pp. 1085–1092. WI 2017, ACM, New York, NY, USA (2017). https://doi.org/10.1145/3106426.3109040
15. Kopeć, W., et al.: VR with older adults: Participatory design of a virtual ATM training simulation. IFAC-PapersOnLine **52**(19), 277–281 (2019). https://doi.org/10.1016/j.ifacol.2019.12.110, http://www.sciencedirect.com/science/article/pii/S2405896319319457. (14th IFAC Symposium on Analysis, Design, and Evaluation of Human Machine Systems HMS 2019)

16. Kowalski, J., et al.: Older adults and voice interaction: a pilot study with google home. In: Extended Abstracts of the 2019 CHI Conference on Human Factors in Computing Systems, pp. 187:1–187:6. CHI EA 2019, ACM, New York, NY, USA (2019). https://doi.org/10.1145/3290607.3312973

17. Martínez-Alcalá, C.I., et al.: Digital inclusion in older adults: a comparison between face-to-face and blended digital literacy workshops. Frontiers in ICT **5**, 21 (2018). https://doi.org/10.3389/fict.2018.00021, https://www.frontiersin.org/article/10.3389/fict.2018.00021

18. McArdle, J.J., Hamagami, F., Meredith, W., Bradway, K.P.: Modeling the dynamic hypotheses of GF-GC theory using longitudinal life-span data. Learn. Individ. Differ. **12**(1), 53–79 (2000)

19. McCormick, M.: Waterfall vs. agile methodology. MPCS, N/A (2012)

20. Mclean, A.: Ethical frontiers of ict and older users: Cultural, pragmatic and ethical issues. Ethics and Inf. Technol. **13**(4), 313–326 (Dec 2011). https://doi.org/10.1007/s10676-011-9276-4, https://doi.org/10.1007/s10676-011-9276-4

21. Skorupska, K.,et al.: All factors should matter! reference checklist for describing research conditions in pursuit of comparable IVR experiments (2021)

22. Skorupska, K., Jaskulska, A., Masłyk, R., Paluch, J., Nielek, R., Kopeć, W.: Older adults' motivation and engagement with diverse crowdsourcing citizen science tasks. In: Ardito, C., et al. (eds.) Human-Computer Interaction - INTERACT 2021, pp. 93–103. Springer International Publishing, Cham (2021)

23. Skorupska, K., Kamil, Warpechowski, Nielek, R., Kopeć, W.: Conversational crowdsourcing for older adults: a Wikipedia chatbot concept (2020)

24. Skorupska, K., Núñez, M., Kopeć, W., Nielek, R.: Older adults and crowdsourcing: Android tv app for evaluating TEDX subtitle quality. Proc. ACM Hum. Comput. Interact. **2**(CSCW), 159:1–159:23 (2018). https://doi.org/10.1145/3274428

25. Skorupska, K., Núñez, M., Kopeć, W., Nielek, R.: A comparative study of younger and older adults' interaction with a crowdsourcing android tv app for detecting errors in TEDX video subtitles. In: Lamas, D., Loizides, F., Nacke, L., Petrie, H., Winckler, M., Zaphiris, P. (eds.) Human-Computer Interaction - INTERACT 2019, pp. 455–464. Springer International Publishing, Cham (2019)

26. Spiekermann, S.: Ethical it innovation: A value-based system design approach, January 2015. https://doi.org/10.1201/b19060

27. Sturdee, M., Alexander, J., Coulton, P., Carpendale, S.: Sketch & the lizard king: Supporting image inclusion in HCI publishing, pp. 1–10, April 2018. https://doi.org/10.1145/3170427.3188408

28. Tai, M.C.T.: Deception and informed consent in social, behavioral, and educational research (SBER). Tzu Chi Med. J. **24**(4), 218–222 (2012). https://doi.org/10.1016/j.tcmj.2012.05.003, https://www.sciencedirect.com/science/article/pii/S1016319012000468

29. Vidoni, M., Cunico, L., Vecchietti, A.: Agile operational research. J. Oper. Res. Soc. **72**(6), 1221–1235 (2021)

30. Vines, J., Pritchard, G., Wright, P., Olivier, P., Brittain, K.: An age-old problem: examining the discourses of ageing in HCI and strategies for future research. ACM Trans. Comput. Hum. Interact. **22**(1), 2:1–2:27 (2015). https://doi.org/10.1145/2696867, http://doi.acm.org/10.1145/2696867

31. De Wit-de Vries, E., Dolfsma, W.A., van der Windt, H.J., Gerkema, M.P.: Knowledge transfer in university-industry research partnerships: a review. J. Technol. Transf. **44**(4), 1236–1255 (2019)
32. Wolfson, N.E., Cavanagh, T.M., Kraiger, K.: Older adults and technology-based instruction: optimizing learning outcomes and transfer. Acad. Manage. Learn. Educ. **13**(1), 26–44 (2014)

Engaging Electricity Users in Italy, Denmark, Spain, and France in Demand-Side Management Solutions

Zbigniew Bohdanowicz$^{(\boxtimes)}$ ⓘ, Jarosław Kowalski ⓘ, and Paweł Kobyliński ⓘ

National Information Processing Institute, Warsaw, Poland
zbigniew.bohdanowicz@opi.org.pl

Abstract. This paper presents the process of uncovering the motivations and barriers for adopting innovative solutions to increase the flexibility of electricity demand among individual consumers. Currently, efforts are being made to decarbonize electricity production with distributed solar and wind renewable energy installations. Such a shift in energy production also requires significant changes on the consumption side, in particular making demand more flexible to match the current situation in the power grid. The challenge in designing demand-side solutions is to accurately identify the needs of individual users so that they are motivated to take advantage of new solutions. Using data from a quantitative survey of electricity consumers in four countries (Italy, Denmark, Spain, France) on energy literacy, values and attitudes towards energy saving and technology, a cluster analysis was carried out which identified five types of electricity users. The segments defined in this way were the basis for conducting qualitative creative workshops with experts dealing with modern solutions in the field of energy and with individual electricity users. Subsequently, this information was supplemented with theoretical knowledge from the field of economic psychology regarding decision making, cognitive processes and motivation. This method allowed, already at the early stage of innovation design, to identify motivations and barriers specific to individual groups of users. The designers of innovative solutions received valuable clues as to how new technologies should be designed in order to ensure that they are well aligned with the habits, needs and rhythm of daily routines of the users.

Keywords: Energy usage flexibility · Demand response · Energy saving · User engagement · Motivations · Smart grid solutions

1 Introduction and Related Works

In this article we describe a method for incorporating a social and user perspective into the process of developing a technological innovation for opening the flexibility of electricity demand among individual energy users, with some preliminary results. The aim of the paper is to communicate this experience to other researchers and developers working on technological innovations, both in the energy sector and elsewhere. We believe that by

© The Author(s) 2022
C. Biele et al. (Eds.): MIDI 2021, LNNS 440, pp. 171–178, 2022.
https://doi.org/10.1007/978-3-031-11432-8_17

using this approach, technological solutions can be better aligned with user needs and preferences, and therefore have a better chance of successful implementation.

In the face of the climate crisis, international efforts are being made to decarbonise the energy sector by developing low-carbon energy sources [1]. Such a change will require a fundamental re-engineering of both the way energy is produced and the way it is used. While changing the way energy is produced is a technological task that requires a plan and a budget, changing the way energy is used implies a social change that requires consumers to understand and accept new ways of using energy.

At present, the decarbonisation of energy in the EU mainly concerns electricity, which in future is to be generated mostly with low emissions technology - alongside nuclear and biomass, wind and solar energy are expected to become important sources. The main problem with wind and solar energy production is its variability and the difficulty in forecasting the available amount. Also, with the development of photovoltaic energy, an increasingly important problem for the power grid is the misalignment of peak energy production (middle of the day) and peak energy consumption (evening hours). This reduces the efficiency of energy production and increases costs. As there is no technology allowing the cost-effective storage of a significant share of the energy consumed, the increasing share of solar and wind power requires individual consumers to adapt their daily rhythm of electricity use to the current availability in the grid [2], so that more renewable energy is used, reducing both costs and greenhouse gas emissions.

Energy consumption patterns of individual consumers affect both the general level of energy consumption and the flexibility of energy demand. Shifting a portion of energy use from peak hours to other times reduces peak load on the grid, allowing energy needs to be met by infrastructure with lower energy production potential. However, the concept of flexible electricity demand is an entirely new concept. Most users are not aware that it is important not only to reduce the amount of energy used but also to adjust consumption patterns to energy availability.

This can be achieved, for example, by avoiding the use of energy-intensive equipment, such as air conditioners, water heaters, or washing machines during peak load hours, or by limiting the intensity of heating or cooling during those periods [3]. Such actions reduce stress on the electricity grid, decrease CO_2 emissions and reduce cost. This is why it is so important not only to develop new technologies that enable a decarbonisation of electricity generation, but also to involve electricity users, as they play a key role in this process.

Designers of systems for smart energy management are confronted with a serious challenge. When optimising the operation of the system, they have to take into account a number of factors, including the needs of energy producers, users, as well as technological possibilities. There are a number of studies highlighting the complexity of this problem and assessing what factors increase the likelihood of success for energy business innovations. Researchers look at this topic from the perspective of business models [4, 5]; assessing how people can be motivated to behave in an environmentally responsible and climate-friendly manner [6] or evaluating price incentives for programmes promoting time shifting of energy consumption [7]. David Halpern highlights the importance of simple actions based on knowledge from behavioural economics to change behaviour and identifies four general characteristics of effective action, summarised under the acronym

EAST (Easy, Attract, Social, Timely) [8]. Researchers also describe the success factors for adapting new technologies for home energy management. Guerassimoff and Thomas explore how a web interface and loyalty programme can keep users engaged in energy management actions [9]. Wilson, Hargreaves and Hauxwell-Baldwin evaluated how perceived risks and benefits affect perception of smart home technologies [10]. Kowalski and Matusiak evaluated how the use of IoT solutions redefines the concept of electricity [11]. This article describes a case study on how best practice and guidance from previous work has been applied to match a user's needs with technological solutions right from the design stage.

2 Methods

In our approach, we decided to use a multi-stage method. This approach was intended to support the design of the user interface of a system for managing household electricity consumption. The following sources were used to gather multidimensional perspective about users and to prepare guidelines on how to modify patterns of their energy consumption behavior:

1. Scientific knowledge on decision making, cognitive processes and motivation. The work included concepts such as Priming, Framing, Social Scripts, Exposure Effect, Heuristics of ease, Intrinsic motivation, Values, Self Identification and others.
2. A quantitative survey of electricity consumers, carried out in June 2020, in the four European countries: France, Spain, Italy and Denmark (N = 3200, random-quota sample, matched to demographic structure in each country, Computer Assisted Web Interview). The aim of the survey was to find out about electricity users in terms of their knowledge, attitudes towards the environment and technology, and their approach to saving energy and its efficient use. The interview lasted on average 20 min.
3. Cluster analysis on data from the quantitative survey, identifying types of electricity users (segmentation, hierarchical clustering method).
4. Qualitative profiles of the typical user (Personas). With this method, the quantitatively defined segments served as a basis for qualitative analysis and preparation of descriptive fictional portraits of characters who are typical representatives of each user segment.
5. Co-design workshops, carried out with both individual electricity users (2 workshops) and experts from the energy sector (1 workshop). Workshops were conducted online, via the freely available Jamboard platform and lasted about 3 h each. The aim of the workshops was to identify the main motivations and barriers to Demand Side Management (DSM) and to develop ways to motivate users to use the two solutions (Fig. 1):

    a. Smart energy management of home appliances
    b. Vehicle-to-Grid charging station at place of work.

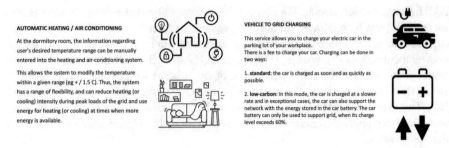

AUTOMATIC HEATING / AIR CONDITIONING

At the dormitory room, the information regarding user's desired temperature range can be manually entered into the heating and air-conditioning system.

This allows the system to modify the temperature within a given range (eg + / 1.5 C). Thus, the system has a range of flexibility, and can reduce heating (or cooling) intensity during peak loads of the grid and use energy for heating (or cooling) at times when more energy is available.

VEHICLE TO GRID CHARGING

This service allows you to charge your electric car in the parking lot of your workplace.
There is a fee to charge your car. Charging can be done in two ways:

1. **standard**: the car is charged as soon and as quickly as possible.

2. **low-carbon**: in this mode, the car is charged at a slower rate and in exceptional cases, the car can also support the network with the energy stored in the car battery. The car battery can only be used to support grid, when its charge level exceeds 60%.

Fig. 1. Concepts evaluated during co-creation workshops.

3   Results

The quantitative survey carried out in the first phase provided knowledge about the structure of electricity users, their level of energy literacy, owned electrical appliances and attitudes. It also evaluated three preliminary concepts of solutions to increase flexibility of electricity demand. This stage showed that most users are open to new solutions concerning energy use and have a positive attitude towards technology. The main motivators for using new solutions were the financial benefits and the reduction of negative environmental impacts.

To better understand the structure of users and to establish a communication strategy tailored to the needs of specific user types, a segmentation analysis (hierarchical clustering method) was carried out on the quantitative survey data, which distinguished 5 user segments, named: Dynamic Traditionalists, Sceptics, Affluent, Supporters, Open and Modest. Figure 2 shows the results in a simplified, synthetic form, the number of dots indicates the relative intensity within the variance between the identified segments.

	Open and Modest 53%	Supporters 20%	Sceptics 13%	Dynamic Traditionalists 12%	Affluent 3%
Energy Saving	●●●	●●●●	●	●●●	●●●●●
Energy Literacy	●●●	●●●●	●	●●	●●●●●
Energy Consumption	●●	●●●	●●	●●●●	●●●●
Main motivation to Save Energy	financial, environmental	financial, environmental	financial	financial, new technology, environmental	financial, environmental, new technology
Energy Equipement	●●	●●●	●	●●●●	●●●●●
Electric / Hybrid cars	●	●●	●	●●●	●●●●●
Age (more dots = older)	●●●●●	●●●	●●●●	●●	●●
Social activity level	●●●	●●●●	●●	●●●●	●●●●
Household size (no. of people)	●●	●●●●	●	●●●●	●●●
Income	●●	●●●	●●●	●●●●	●●●●●
General Values	Sensitive, cooperative, fulfilled	Sensitive, energetic, positive, fulfilled	Traditional, sceptical, reserved	Traditional, reserved, active, dynamic	Active, ambitious, open-minded, like to stand out
Environmental Values	●●●●	●●●●	●●	●●	●●●
Attitude to technology	●●●	●●●●	●●	●●●●	●●●●●
Education	●●●	●●●●	●●	●●●●	●●●●●
City size	●●	●●●	●●	●●●●	●●●●●

Fig. 2. Summary results of the segmentation analysis performed on the data collected in the quantitative survey of electricity users (N = 3200), showing the dimensions that differentiate the different user groups.

This analysis has helped to identify priorities in communication with users, tailored to their characteristics. It also showed that those individuals who are most sceptical of change and reluctant to save energy also have the lowest relative income and, regardless of their views, already use significantly less energy than users from the other segments. The analysis showed that users in the other segments are generally open to changing the way they use energy and have a positive attitude towards technology, which indicates a relatively high potential for acceptance of new solutions.

Once the key user groups were identified, qualitative profiles of the typical user (personas) were developed so that the characteristics of a typical user representing a segment could be portrayed in an accessible way. Figure 3 shows an example of such a description, for a user from the Open and Modest segment.

Fig. 3. Qualitative description (persona) of a fictional energy user, representing the Open and Modest segment.

The personas prepared in this way were used in co-design workshops with energy experts and individual users. The results of the workshop indicated a number of non-obvious, previously unnoticed needs and motivations. For example, in the case of Open and Modest users, the relatively oldest user group, it emerged that an important motivation for using smart energy management devices may be a dimension not directly related to energy consumption, namely - safety. This dimension manifests itself in the need for physical safety - e.g. automatically controlled electrical appliances can reduce the risk of fire caused by burning electrical appliances that have not been switched off by mistake. An energy management system can prevent such risks with built-in functions that warn the user about devices that have not been switched off. The safety dimension can also be understood as support in the use of electrical appliances through automation of controls. It is difficult for an older person to manually set the operating parameters of a number of electrical appliances and here the DSM system can help by automating their management. Finally, there is also a financial aspect to the safety dimension, because by reducing the costs associated with energy use, financial security is increased, which is particularly important for senior users who expect their income to decline with age.

The workshop also brought new insights for users from other segments regarding their motivation to use smart electricity management systems, that due to space constraints are not presented here.

4 Discussion

The above results were complemented with scientific knowledge on decision making, cognitive processes and motivation. On this basis, guidelines have been developed on how to use knowledge from both the scientific research and from the users to prepare a system that involves users to make changes in the way they use electricity, in order to reduce the load on the electricity grid and thus reduce the costs of energy production and associated greenhouse gas emissions. Getting users to change their behaviour is a difficult task, as it requires the modification of established beliefs and habits. It also requires additional effort. Therefore, actions encouraging people to change their current pattern of energy use should be multi-dimensional, so as to show the importance of the values on which the new behaviour is based, strengthen the motivation to maintain the desired behaviour, provide knowledge justifying the change of behaviour and activate actions on a broader than individual level. A good example of an environment where such programmes can be pilot tested is a university campus. A university can promote values and behaviours through its activities, organisation and communication to students. The effectiveness of such activities is enhanced by the fact that the university creates a cohesive and well-defined community that is open to change and the introduction of new knowledge-based solutions.

5 Conclusions

With the method presented, it was possible, already at the early stage of designing innovations, to identify motivations and barriers specific to particular groups of users. The developers of innovative solutions received a detailed report, with valuable clues on how to integrate new technologies with the habits, needs and rhythm of daily activities of users.

The process described proved to be simple to implement and cost-effective. Although it required additional work for the researchers - analysis of quantitative data to prepare consumer segments and personas, or the work needed to organise and conduct workshops, however, this did not involve a significant increase in total costs for the project. Work on the social aspects of innovations for energy management was carried out in parallel with the development of the technology and the results were discussed with engineers and technical experts on an ongoing basis. Although difficult to quantify at this stage, it seems reasonable to argue that a better understanding of the social aspects of the application of new technologies saves significant resources and time at the design stage and increases the chances of successful implementation. The approach presented can also inspire designers of other innovations. When designing new solutions, getting to know users well and matching their needs and technical capabilities is essential for success. We believe that this practical example of how the user perspective can be incorporated into the process of designing technological solutions can be useful for other design teams looking for technological solutions in various fields.

The presented approach also has some limitations. Only the first stages of the innovation co-creation process have been presented here, consisting of collecting information about users and evaluating concepts. Currently, prototypes of solutions for managing household electricity consumption are being designed, taking into account the knowledge gathered in the process presented here. The evaluation of these prototypes, and then of the finished solutions, will be a true test of the validity of the approach presented.

Acknowledgements. This work emanated from research conducted with the financial support of the European Commission through the H2020 project, ebalance+ (Grant Agreement 864283).

References

1. Masson-Delmotte, V., et al.: IPCC Sixth Assessment Report (AR6), WG1. Cambridge University Press, Cambridge (2021)
2. Jamasb, T., Llorca, M.: Energy systems integration: economics of a new paradigm. Econ. Energy Environ. Policy **8**, 7–28 (2019). https://doi.org/10.5547/2160-5890.8.2.tjam
3. Jones, R.V., Fuertes, A., Lomas, K.J.: The socio-economic, dwelling and appliance related factors affecting electricity consumption in domestic buildings. Renew. Sustain. Energy Rev. **43**, 901–917 (2015). https://doi.org/10.1016/j.rser.2014.11.084
4. Brown, D., Hall, S., Davis, M.E.: Prosumers in the post subsidy era: an exploration of new prosumer business models in the UK. Energy Policy **135**, 110984 (2019). https://doi.org/10.1016/j.enpol.2019.110984
5. Warren, P.: Demand-side policy: mechanisms for success and failure. Econ. Energy Environ. Policy **8** (2019). https://doi.org/10.5547/2160-5890.8.1.pwar
6. Verbong, G.P.J., Beemsterboer, S., Sengers, F.: Smart grids or smart users? Involving users in developing a low carbon electricity economy. Energy Policy **52**, 117–125 (2013). https://doi.org/10.1016/j.enpol.2012.05.003
7. Christensen, T.H., et al.: The role of competences, engagement, and devices in configuring the impact of prices in energy demand response: findings from three smart energy pilots with households. Energy Policy **137**, 111142 (2020). https://doi.org/10.1016/j.enpol.2019.111142
8. Halpern, D.: Inside the Nudge Unit: How Small Changes Can Make a Big Difference. WH Allen (2016)
9. Guerassimoff, G., Thomas, J.: Enhancing energy efficiency and technical and marketing tools to change people's habits in the long-term. Energy Build. **104**, 14–24 (2015). https://doi.org/10.1016/j.enbuild.2015.06.080
10. Wilson, C., Hargreaves, T., Hauxwell-Baldwin, R.: Benefits and risks of smart home technologies. Energy Policy **103**, 72–83 (2017). https://doi.org/10.1016/j.enpol.2016.12.047
11. Kowalski, J., Matusiak, B.E.: End users' motivations as a key for the adoption of the home energy management system. J. Manag. Econ. **55**(1), 13–24 (2019)

Remote Scientific Conferences After the COVID-19 Pandemic: The Need for Socialization Drives Preferences for Virtual Reality Meetings

Agata Kopacz[ID], Anna Knapińska[✉][ID], Adam Müller, Grzegorz Banerski[ID], and Zbigniew Bohdanowicz[ID]

National Information Processing Institute, Warsaw, Poland
{agata.kopacz,anna.knapinska,adam.muller,grzegorz.banerski,
zbigniew.bohdanowicz}@opi.org.pl

Abstract. The COVID-19 pandemic continues to exert influence on the scientific community: circumstances have forced academics to engage more frequently in technology-mediated activities, including their participation in remote and virtual conferences. In this article, we contemplate immersive virtual environments: we verify researchers' motivations and constraints in the context of online conferences, and discover in what elements of such conferences researchers wish to participate in virtual reality (VR). A survey was administered using a computer-assisted web interview (CAWI) questionnaire among the sample of 1,575 academics with the POL-on database as the sampling frame. The results indicate that individuals' contrasting needs and attitudes toward technology determine the degree to which they look favourably upon both remote conferencing and VR. Immersive virtual environments appear to satisfy the need for socialization; ordinary remote conferences fulfil the need for security and the achievement of fundamental conference goals, such as establishing collaboration and publishing research results. Conferences that are hosted remotely must be relevant to the needs of researchers and meet their discrete expectations; only then will such events prove valuable enough that researchers are willing to continue participating in them after the pandemic subsides.

Keywords: Remote scientific conference · Virtual reality meetings · COVID-19 pandemic · Need for socialization · Attitudes toward technology-mediated activities

1 Introduction

Conferences are one of the primary means by which new scientific knowledge is disseminated and that researchers interact with their peers [10, 17]. Prior to 2020, multi-day, all-day, in-person conferences were regarded as the norm. During the COVID-19 pandemic, an abundance of opportunities has arisen for researchers to participate in online

© The Author(s) 2022
C. Biele et al. (Eds.): MIDI 2021, LNNS 440, pp. 179–188, 2022.
https://doi.org/10.1007/978-3-031-11432-8_18

scientific events for shorter durations, while balancing their everyday duties from the comfort of their homes or labs. This has also presented challenges for those who have unsuitable circumstances to participate in such events remotely, and for those who feel uncomfortable with the less immersive interactions of such events. The second group benefits from their attendance at in-person conferences, and virtual reality represents the closest alternative. In this article, we focus on the use of immersive virtual environments at remote conferences. We examine which groups exhibited the strongest desire to use the technology, and tested researchers' eagerness to participate in particular conference elements that are delivered in virtual reality (VR).

2 Literature Review

By attending prominent scientific conferences, academics gain visibility and recognition: a fundamental "currency" in academia. They can also contribute to the maintenance of established order [18], as well as reinforcing the Matthew effect [12]; that is, the accumulation of scientific advantages by the scholars who already possess them. Such events entailed limited numbers of attendees, and were often difficult for those attendees to reach due to financial or time constraints [7, 11].

During the COVID-19 pandemic, the importance of online and virtual conferences has increased markedly [1, 19]. This format has enhanced the accessibility of conferences to previously underrepresented groups – including female, minority, and early-career researchers [2, 3, 8, 16, 19]. Moreover, the environmental virtues of remote conferences—chiefly their reduced carbon footprint—have been lauded [1, 5, 8]. The format, however, also entails a number of disadvantages: attendees are no longer physically present at the venues; speakers are largely deprived of the opportunity to observe their audience's reactions; distractions are frequent; and such events are exposed to continual risk of technical difficulties. All of the above might prove significant impediments to mutual interaction [18]. Networking, which is widely considered a key benefit of conference attendance among researchers, can be also disrupted in remote environments [1]. Scientific collaborations are frequently initiated during backstage chats and coffee breaks. Although organizers of remote conferences endeavour to provide space for informal conversations, "you cannot enjoy a virtual drink!" [15, p. 3].

During traditional, in-person conferences, exclusive pools of scientists are immersed fully in reality; such events do not necessitate the extensive use of electronic devices. Although remote online events are more inclusive and bring together a larger number of attendees from a wider variety of regions [19], those attendees can see each other's faces only on their screens, while they remain at their places of residence or work. One alternative, and a chance to avoid "Zoom fatigue" [14], involves the delivery of conferences in VR [6]. Pioneering events that utilized the technology were hosted in Second Life, an application released in 2003 that allows users to create their own avatars and interact in a computer-simulated world [4, 13]. Standalone virtual platforms tailored to scientific conferences requirements have also been developed [20, 21]. Presently, Mozilla Hubs, a social web-based virtual space is gaining in popularity [1, 6].

In line with research, several strengths and weaknesses of VR conferences can be identified. An evaluation study of a gathering organized by the IBM research team in the Second Life environment revealed the participants' overall satisfaction; attendees found poster sessions to be the most successful aspect of the conference [4]. A pilot experiment of Merck Research Laboratories achieved similar results: 94% of the participants found the virtual conference to be valuable, and 59% declared that the virtual poster session was more effective than a face-to-face one [21]. Virtual reality has the potential to overcome one of the most significant drawbacks of online conferences: reduced networking opportunities [1]. During the IEEE Conference on Virtual Reality and 3D User Interfaces in 2020 (IEEEVR2020), participants were offered to participate using three different platforms: Twitch, (an interactive livestreaming service); Slack (a communication platform); and Mozilla Hubs (a VR platform). Although attendees reported that Mozilla Hubs was the least effective for socializing and building networks, the platform's users perceived social presence was the highest of the three options; its users described the social gatherings as "fun and playful" [1, p. 9]. The most serious disadvantage of VR conferences is the presence of technical and technological obstacles; unstable internet speed [9] or the discomfort of wearing a VR headset for long periods [1] carry the potential to create frustration. Primarily for this reason, many IEEEVR2020 participants fled the VR environment for simple video streams. Importantly, this observation also held true for individuals who had previous VR experience [1].

3   Methodology, Sample, and Respondents

A survey was conducted by the National Information Processing Institute (OPI PIB) in the form of a computer-assisted web interview (CAWI) questionnaire between 28 January and 8 February 2021. The sample size was 1,575. The POL-on database of academics in Poland (https://polon.nauka.gov.pl/siec-polon) was used as the sampling frame. At the outset, the sample of academics was drawn; then, they received an email that contained an invitation to participate in the survey. The sampling scheme considered the distribution of characteristics in the population of academics in Poland in terms of gender, degree or title, type of institution, and scientific field (random quota sampling).

We assessed participation in remote scientific conferences in three primary aspects: a) attitudes toward participation in remote scientific conferences – drivers and barriers; b) attitudes toward participation in remote scientific conferences in the form of meetings in immersive virtual reality (IVR) – drivers and barriers; and c) preferences regarding the format of remote scientific conferences – including speaking time, methods of presenting results, and other events.

We divided the respondents using two criteria: overall willingness to participate in remote scientific conferences after the COVID-19 pandemic, and willingness to participate in remote scientific conferences in VR. In both categories, respondents were asked to indicate whether they were "for" or "against" participating in these event types. In this way, we obtained four subsamples, which we compared in terms of their motivations and preferences. Table 1 presents the split of the respondents and the sizes of the groups.

Table 1. Split of respondents and group sizes.

		Willingness to participate in remote conferencing in VR	
		Yes	No
Willingness to participate in remote conferencing after the COVID-19 pandemic*	Yes	A: (n = 244)	B: (n = 470)
	No	C: (n = 193)	D: (n = 324)

Respondents who answered, "I don't know" (n = 344) were excluded from the analyses.

The following abbreviations are used in the remainder of this article: VR_Yes_Remote_Yes (group A); VR_No_Remote_Yes (group B); VR_Yes_Remote_No (group C); VR_No_Remote_No (group D).

The respondents also provided open-ended qualitative responses in which they indicated the reasons for their preferences regarding events that utilize videoconferencing or VR technology. We then categorised the motivations they mentioned. The content of the open-ended questions was later quantitatively coded to the motivation categories.

4 Results

The survey considered respondents' general willingness to participate in remote conferences, and that the potential influence of current epidemiological circumstances. Proponents of postpandemic remote conferencing—including those interested (group A – 93%) and those uninterested (group B – 95%) in using VR as part of the conference experience—were significantly more likely to state they would also attend such events during the pandemic. Among those who wished not to participate in remote conferences after the pandemic, 41% (group C) would consider doing so in the current circumstances; these are the same respondents who exhibited interest in experiencing conferences in VR. Group D comprises respondents who were interested in neither postpandemic remote conferencing nor VR technology, but considered participating in remote conferences during the pandemic (62%).

The sources of the respondents' unwillingness to attend remote conferences are compelling when considered from the perspectives of VR proponents and skeptics. In the VR proponents group, those uninterested in postpandemic remote conferencing (group C) were significantly more likely (85%) to cite 'difficulty interacting with other participants' compared to those who considered attending such events (group A – 56%). Conversely, among the VR skeptics, those reluctant to participate in remote conferences (group D) were significantly more likely (54%) than online conference supporters (group B; 23%) to state that they were discouraged by the 'lower status of such conferences'.

With regard to respondents' motives for participating in remote conferences, a similarity can be observed in the distributions of results across groups A and B, and across groups C and D. The overall desire to attend remote conferences in groups A and B is reflected in the number of motives indicated. The key reasons stated among groups A and B are presented in Table 2.

Members of groups C and D were significantly less likely to indicate such motives; this is consistent with their reluctance to continue attending remote conferences after the pandemic. Nevertheless, we observed greater desire to meet the primary goals of conferences among members of group D compared to group C (Table 2). Interestingly, group B is significantly more likely than group C to attach high value to the practical aspects of such events.

Table 2. What would encourage you to participate in a remote conference?

	A:VR_Yes Remote_Yes	**B:**VR_No Remote_Yes	**C:**VR_Yes Remote_No	**D:**VR_No Remote_No
Conference subjects	72%^{CD*}	78%^{CD}	52%^{ABD}	68%^{ABC}
Personal safety during the COVID-19 pandemic	65%^{CD}	60%^{CD}	25%^{AB}	30%^{AB}
Participation of compelling and well-known speakers	63%^{CD}	57%^{CD}	38%^{AB}	41%^{AB}
Opportunities to identify current scientific trends	60%^{CD}	65%^{CD}	33%^{ABD}	48%^{ABC}
Financial considerations and cost containment	58%^{CD}	57%^{CD}	20%^{AB}	24%^{AB}
No need to travel to the conference	55%^{CD}	56%^{CD}	9%^{AB}	14%^{AB}
Opportunities to present research personally	49%^{CD}	46%^{CD}	19%^{ABD}	32%^{ABC}
Opportunities to learn about othersí projects	42%^{CD}	34%^{CD}	16%^{ABD}	24%^{ABC}
Collecting comments from others on my own research	35%^{CD}	39%^{CD}	18%^{AB}	24%^{AB}
Ecology - no need to travel	31%^{CD}	29%^{CD}	6%^{AB}	8%^{AB}
Convenient hours for sessions	23%^{CD}	24%^{CD}	11%^{AB}	12%^{AB}
Opportunities to maintain work-life balance	22%^{CD}	24%^{CD}	8%^{AB}	6%^{AB}
Convenient conference dates	19%^{CD}	19%^{CD}	11%^{AB}	11%^{AB}

The letters in superscript denote that the value differs significantly from that in the column denoted by that letter. Statistical differences were computed as the result of the bilateral equality test for column proportions at the 95% significance level. Green denotes the highest score; red denotes the lowest.

We also asked respondents what specific events they wished to attend during remote conferences (Table 3). Attention among those who were open to using VR technology (groups A and C) focused more than the VR skeptics (groups B and D) did on networking. Unsurprisingly, formal aspects of the conference were of higher importance among group B than group C.

Table 3. Which specific events at a remotely hosted conference would you wish to attend?

	A:VR_Yes Remote_Yes	**B:**VR_No Remote_Yes	**C:**VR_Yes Remote_No	**D:**VR_No Remote_No
Live presentation sessions	80%	82%^{C}	69%^{B}	79%
Informal meetings that enable networking	41%^{BD}	30%^{AC}	44%^{BD}	29%^{AC}
Keynote speeches	32%^{C}	35%^{C}	19%^{ABD}	32%^{C}

The letters in superscript denote that the value differs significantly from that in the column denoted by that letter. Statistical differences were computed as the result of the bilateral equality test for column proportions at the 95% significance level. Green denotes the highest score; red denotes the lowest.

The results differed when we asked the respondents about their willingness to participate in conferences organized in VR (Table 4). The respondents who indicated openness
to VR conferencing tools and simultaneous reluctance to participate in standard remote
conferencing (group C) were interested in other, more social events than members of
group B. When in which activity types they wished to participate during VR conferences,
the technophiles (group A) exhibited significantly more interest than other groups in each
of the types proposed: from formal activities, through informal ones, to cultural ones. In
contrast lie the techno-skeptics (group D), who exhibited significantly less interest than
the others groups in each of the activity types. In three dimensions, however, they do not
differ from group B, with whom they share an aversion to VR. Both groups also exhibited
interest in informal aspects of VR conferences to similarly low degrees. This explains
the reluctance of the groups to engage with VR – a technology that offers opportunities
to experience closer contact with others.

Table 4. Which specific events at a remotely hosted VR conference would you wish to attend?

	A:VR_Yes Remote_Yes	B:VR_No Remote_Yes	C:VR_Yes Remote_No	D:VR_No Remote_No
Live presentation sessions	73%[BCD]	54%[AD]	61%[AD]	43%[ABC]
Poster sessions in large VR rooms	63%[BCD]	48%[AD]	52%[AD]	33%[ABC]
Informal meetings that enable networking	59%[BCD]	40%[AD]	48%[AD]	30%[ABC]
Conversations with others in the form of prearranged one-on-one meetings	53%[BD]	29%[AC]	46%[BD]	23%[AC]
Keynote speeches	51%[BCD]	33%[AD]	39%[AD]	23%[ABC]
Coffee breaks that enable spontaneous meetings with other conference participants	50%[BCD]	23%[AC]	40%[ABD]	19%[AC]
Virtual tours of host cities/countries	45%[CD]	42%[CD]	27%[AB]	24%[AB]
Virtual cultural events, such as concerts	43%[BCD]	34%[AD]	31%[AD]	20%[ABC]
Cocktail parties in the form of large VR meetings	27%[BD]	11%[AC]	24%[BD]	7%[AC]

*The letters in superscript denote that the value differs significantly from that in the column denoted by that letter. Statistical
differences were computed as the result of the bilateral equality test for column proportions at the 95% significance level. Green
denotes the highest score; red denotes the lowest.

Finally, we more closely examine the characteristics of the members of each group.
We observed a similar pattern of results in groups A (VR_Yes_Remote_Yes) and B
(VR_No_Remote_Yes) when compared with groups C (VR_Yes_Remote_No) and D
(VR_No_Remote_No). Individuals who exhibited interest in participating in postpandemic remote conferences—both those interested and those uninterested in using VR
technology—had significantly more experience with online conferencing (Fig. 1). We
can conclude from this that the experience of participating in remote conferences
determines respondents' willingness to participate in such events in the future.

Moreover, we observed age differences between group A and the others (Fig. 2).
Group A, which exhibited positive attitudes toward both remote and VR conferencing,
is dominated by those aged under 45.

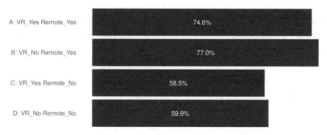

Fig. 1. Experience with online conferencing

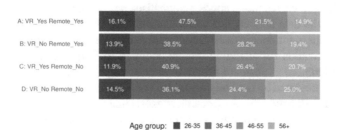

Fig. 2. Age distribution

Interestingly, group C with individuals interested in using VR technology, but uninterested in attending remote conferences per se largely comprise males (Fig. 3). Compared to this group, three times as many female members of group B stated that they were considering participating in online conferences, but did not wish to use VR. This pattern might indicate more pragmatism among the female group members than among the male ones.

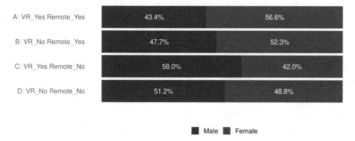

Fig. 3. Gender distribution

Additionally, we coded an open-ended question in which survey participants could elaborate on their reasons for preferring VR or classic remote conferences. Among the proponents of classic remote conference, the most common reasons included unfamiliarity with VR technology and the desire to use familiar solutions, aversion to VR, lack of added value, and a mismatch between the novelty of the format and the prestige of academic conferences. Some respondents believed that remote conference contact is closer to that experienced during in-person meetings. Analogously, proponents of VR indicated superior imitation of real conference environments, contact with other participants, and curiosity about the new technology.

5   Conclusions

The two primary variables we considered—attitudes toward remote conferencing and toward virtual reality—paint a more accurate picture of researchers, whose openness to technology varies in the context of scientific conferences. We established four discrete groups: technophiles, techno-skeptics, pragmatists, and socializers.

Based on the data collected, it seems that the techno-skeptics are the most likely to continue with traditional in-person conferences that are not mediated by technology. Their key motivation for attending remote conferences is to accomplish the primary goals of conferences as standard academic events; they pay little attention to the new opportunities presented by remotely organized conferences, beyond personal safety during the COVID-19 pandemic. As research suggests that the most rewarding elements of participation in research conferences are related to "attending", "being seen", and "seeing others" [17], we may assume that the techno-skeptics consider physical presence to influence the "flow of academic understanding" [10, p. 28]. The respondents' seniority correlates with their treatment of remote conferences as lower-ranked events that are unable to enhance participants' recognition in their fields.

The conclusions differ among VR-skeptics who exhibit positive attitudes toward remote conferencing – the so-called pragmatists that see opportunities to save time and reduce costs without compromising the goals usually pursued at conferences. The pragmatic attitudes of women are particularly noteworthy. This may corroborate the accessibility of remote conferences to a wider pool of participants – including those who have care duties [3]. Perhaps these additional responsibilities explain the pragmatists' skepticism toward VR – a technology that entails "full immersion"; when participation in conferences happens alongside child- or elderly care, immersion can act as a hindrance rather than an advantage.

The technophiles, like the pragmatists, were open both to remote conferencing and to VR technology. This group is characterized by its exceptional motivation to participate in conferences organized online; it is also the group most willing to participate in almost all conference events organized in VR. Its members are relatively young, which can inform their positive attitudes toward VR and their desire for new opportunities to participate in events. This might encourage organizers of online scientific conferences for early-career scientists to offer attendees unique value propositions rather than only the standard agenda items [1]. This serves as another acknowledgement of remote conferences' greater inclusiveness.

Among the last group we distinguished—socializers who have little desire to participate in online conferences—VR seems to be an opportunity to experience closer relations than at traditional remote conferences. Conversations and informal meetings with others, whether during online or VR-mediated events, matter more to this group than the others. They stand in opposition to the pragmatists, who pay much less attention to these aspects. Those who focus on contact with others may be the most acutely affected by

"Zoom fatigue" [14], and hope to overcome it in VR. Since socialization and networking are the strongest predictors of the overall satisfaction participants derive from remote conferences [1], this should inform the development of future events. Perhaps, as Ahn et al. [1] suggest, strictly scientific points in conference agendas—such as Q&A sessions and keynote speeches—should be arranged on live streaming platforms; this may transfer socializing activities to the virtual world more effectively. Since researchers who attend VR conferences are keen to repeat such experiences [21], combining streaming platforms with immersive environments seems an effective solution both during and after the COVID-19 pandemic.

References

1. Ahn, S.J., Levy, L., Eden, A., Won, A.S., MacIntyre, B., Johnsen, K.: IEEEVR2020: exploring the first steps toward standalone virtual conferences. Front. Virtual Real. **2**, 648575 (2021). https://doi.org/10.3389/frvir.2021.648575
2. Biggs, J., Hawley, P.H., Biernat, M.: The academic conference as a chilly climate for women: effects of gender representation on experiences of sexism, coping responses, and career intentions. Sex Roles **78**(5–6), 394–408 (2017). https://doi.org/10.1007/s11199-017-0800-9
3. Black, A.L., Crimmins, G., Dwyer, R., Lister, V.: Engendering belonging: thoughtful gatherings with/in online and virtual spaces. Gend. Educ. **32**(1), 115–129 (2020). https://doi.org/10.1080/09540253.2019.1680808
4. Erickson, T., Shami, N.S., Kellogg, W.A., Levine, D.W.: Synchronous interaction among hundreds: an evaluation of a conference in an avatar-based virtual environment. In: CHI 2011: Proceedings of the SIGCHI Conference on Human Factors in Computing Systems, p. 503–512. Association for Computing Machinery, New York (2011). https://doi.org/10.1145/1978942.1979013
5. Fraser, H., Soanes, K., Jones, S.A., Jones, C.S., Malishev, M.: The value of virtual conferencing for ecology and conservation. Biol. Conserv **31**(3), 540–546 (2017). https://doi.org/10.1111/cobi.12837
6. Fuller, E.G.: WIP: Mozilla hubs classes fight feelings of isolation and online fatigue. Paper presented at 2021 ASEE Virtual Annual Conference Content Access, Virtual Conference (2021). https://peer.asee.org/38092
7. Gichora, N.N., et al.: Ten simple rules for organizing a virtual conference – anywhere. PLoS Comput. Biol. **6**(2), e1000650 (2010). https://doi.org/10.1371/journal.pcbi.1000650
8. Johnson, R., Fiscutean, A., Mangul, S.: Refining the conference experience for junior scientists in the wake of climate change (2020). https://arxiv.org/abs/2002.12268
9. Le, D.A., MacIntyre, B., Outlaw, J.: Enhancing the experience of virtual conferences in social virtual environments. In: 2020 IEEE Conference on Virtual Reality and 3D User Interfaces Abstracts and Workshops (VRW), pp. 485–494 (2020). https://doi.org/10.1109/VRW50115.2020.00101
10. de Leon, F.L.L., McQuillin, B.: The role of conferences on the pathway to academic impact. J. Hum. Resour. **55**(1), 164–193 (2018). https://doi.org/10.3368/jhr.55.1.1116-8387R
11. Mair, J., Lockstone-Binney, L., Whitelaw, P.A.: The motives and barriers of association conference attendance: evidence from an Australasian tourism and hospitality academic conference. J. Hosp. Manag. Tourism **34**, 58–65 (2018). https://doi.org/10.1016/j.jhtm.2017.11.004
12. Merton, R.K.: The Matthew effect in science: the reward and communication systems of science are considered. Science **159**(3810), 56–63 (1968). https://doi.org/10.1126/science.159.3810.56

13. Messinger, P.R., et al.: Virtual worlds past, present, and future: new directions in social computing. Decis. Support Syst. **47**, 204–228 (2009). https://doi.org/10.1016/j.dss.2009.02.014

14. Nadler, R.: Understanding "Zoom fatigue": theorizing spatial dynamics as third skins in computer-mediated communication. Comput. Compos. **58**, 102613 (2020). https://doi.org/10.1016/j.compcom.2020.102613

15. Oester, S., Cigliano, J.A., Hind-Ozan, E.J., Parsons, E.C.M.: Why conferences matter – an illustration from the international marine conservation congress. Front. Mar. Sci. **4**, 257 (2017). https://doi.org/10.3389/fmars.2017.00257

16. Raby, C.L., Madden, J.R.: Moving academic conferences online: aids and barriers to delegate participation. Ecol. Evol. **11**, 3646–3655 (2021). https://doi.org/10.1002/ece3.7376

17. Rowe, N.: "When you get what you want, but not what you need": the motivations, affordances and shortcomings of attending academic/scientific conferences. IJRES **4**(2), 714–729 (2018). https://doi.org/10.21890/ijres.438394

18. Sá, M.J., Ferreira, C.M., Serpa, S.: Virtual and face-to-face academic conferences: comparison and potentials. J. Educ. Soc. **9**(2), 35–47 (2019). https://doi.org/10.2478/jesr-2019-0011

19. Sarabipour, S.: Virtual conferences raise standards for accessibility and interactions. eLife **9**, e62668 (2020). https://doi.org/10.7554/eLife.62668

20. Shirmohammadi, S., Hu, S.Y., Ooi, W.T., Schiele, G., Wacker, A.: Mixing virtual and physical participation: the future of conference attendance? In: 2012 IEEE International Workshop on Haptic Audio Visual Environments and Games (HAVE 2012) Proceedings. IEEE: Munich, pp. 150–155 (2012). https://doi.org/10.1109/HAVE.2012.637445

21. Welch, C.J., Fare, T.: Virtual conferences becoming a reality. Nat. Chem **2**, 148–152 (2010). https://doi.org/10.1038/nchem.556

Sometimes It's Just a Game: The Pros and Cons of Using Virtual Environments in Social Influence Research

Grzegorz Pochwatko[1]([✉])(ID), Justyna Świdrak[2](ID), and Dariusz Doliński[1](ID)

[1] Institute of Psychology, Polish Academy of Sciences, Warsaw, Poland
gp@psych.pan.pl
[2] SWPS University of Social Sciences and Humanities,
Faculty of Psychology in Wroclaw, Wroclaw, Poland
https://psych.pan.pl

Abstract. Classic social influence effects are present both in games and virtual environments, similarly to real life. The use of games and virtual environments to study them offers the possibility to better control the experimental situation but also brings limitations. On one hand, sequential request techniques of social influence are studied in virtual environments, which enables the control of the experimental situation at the laboratory level. On the other hand, mere presence in the laboratory, devices for measuring physiological responses and awareness of participation in the game provide additional confounding variables that influence the results. We show examples of successful and unsuccessful replications of the foot-in-the-door, door-in-the-face and foot-in-the-face effects accompanied by the analysis of the indicators of physiological arousal. Virtual environments are useful tools for social psychology, but they need to be applied carefully because even a serious game is sometimes just a game.

Keywords: Virtual environments · Social influence ·
Foot-in-the-door · Door-in-the-face · Foot-in-the-face

1 Theoretical Context

1.1 Serious Virtual Environments

VR and AR systems are becoming an integral part of our world. It is no longer just about research environments, computer games or experimental interfaces. Like massively multiplayer online games and social networking sites in the last three decades, nowadays virtual reality develops rapidly driven by the users' creativity and collaborations. We are dealing with mass social environments, entertainment and training applications, immersive apps improving the health and well-being of users, or standard applications with new, 3D interfaces. The growing number of users of virtual environments (VE) means that they more and

C. Biele et al. (Eds.): MIDI 2021, LNNS 440, pp. 189–197, 2022.
https://doi.org/10.1007/978-3-031-11432-8_19

more often contact each other through these platforms where they also meet their inhabitants - artificial agents of various forms. Representations of VR users are most often (but not necessarily) virtual humans, or at least humanoids (especially in the case of some environments that have a simplified, even cartoonish form). Also, virtual agents often take the humanoid form - in this case, we deal with artificial virtual humans. Advances in technology have made it sometimes difficult to recognize whether we deal with a human or an algorithm shaped as one. The VE users see in front of them just a virtual character with which they can interact, as in the real world. Even relatively simple computer interfaces can cause users to behave in a manner characteristic of interaction with another human being, for example prompting them to apply rules of politeness [13,14]. When interacting with (artificial) virtual humans in VE, this is even easier to observe due to the variety of possible behaviours that can be simulated [11], which stands in line with the media equation theory [17] and the media richness theory [4]. Research on social influence and compliance to requests demonstrated that virtual humans in VEs influence real people. We know that people comply in VE as they do in real life, but the authors devote less attention to the mechanisms of this influence and the possible consequences. In this context, it is also worth paying attention to the development of augmented reality (AR), in which virtual elements are mixed with the real world and robotics, in particular humanoid social robots that slowly enter physical reality in the form of surrogates (which can be considered a physical equivalent to avatars in VR) and autonomous robots (analogues of agents in VR).

1.2   Social Influence Techniques

Social influence is understood as changing the attitudes, perceptions, emotions, and behaviour of others [1,2]. Many classic experiments have been replicated in VE. For example, a) social facilitation/inhibition effects, that is, improving the performance of simple tasks and worsening the performance of complex tasks in the presence of virtual humans [15,18]; b) effectiveness of persuasion (increased by the level of realism of the virtual human and gender compatibility with user) [10]; c) conformity (giving answers in line with virtual agents, i.e. giving incorrect responses more often when agents also gave ones compared to the situation in which agents gave correct responses) [12].

Most often, social influence related phenomena are shown through influence techniques [6]. Research on discreet attempts to influence others originated from works of Freedman and Fraser [9], who hypothesised that people who agree to fulfill a small request would be more likely to comply with a more difficult request. In the original experiments, people who agreed to sign a petition to keep California clean, to improve road safety, or agreed to put a small label in the window encouraging compliance with these petitions, were more likely than in the control group to agree to put up an ugly sign in front of their house with the words "drive carefully". The authors named this foot-in-the-door technique (FITD). Another technique, assuming the reverse order of formulated demands proposed by Cialdini [3], is the so-called door-in-the-face (DITF). Its typical course is as follows: making a large, difficult request followed by a refusal that prompts people to comply if the requester withdraws the first one and makes another, much easier request. In the original study, participants were asked to give their consent to act as guardians of juvenile offenders, which required two hours a week for two years (the vast majority of participants refused to comply with this request). The smaller request that followed was to help during a two-hour trip with a group of those young people to the zoo. Participants who previously refused to fulfil a larger request were more likely to agree to a smaller one than in the control group.

Since both agreement and refusal to comply with the first request may, under certain conditions, lead to compliance, the third technique, which is a compilation of the FITD and DITF, was proposed. The foot-in-the-face technique (FITF) proposed by Doliński [5] also assumes the sequential formulation of two requests, this time with the same degree of difficulty. The response to the first request paves the way for the effectiveness of this technique. Those who agree to fulfil the first request are on the "agreement route" (as in the FITD), and those who refuse to comply with the first request are on the "refusal route" (as in the DITF).

Field experiments, used in classic research on the abovementioned techniques, make it impossible to accurately trace the process that takes place at the time when subsequent requests are formulated. Whereas conducting laboratory experiments allows both the use of repeated measures of self-report variables, continuous observation of behaviour and measurement of psychophysiological variables (e.g. [8]). Research in VE combine a high level of ecological validity of field studies with the precision of control and measurement offered by a laboratory setting. The use of VE makes it possible to recreate the natural behaviour of participants under controlled conditions, including the influence of personality traits, distortions resulting from the appearance and behaviour of avatars, both of the user and partners of social interaction. In addition, it is possible to continuously and precisely record the behaviour of participants (position in space, position in relation to other objects, motor skills) and to record psychophysiological variables (such as electrodermal activity, heart rate variability). Unlike field experiments, it is possible to reproduce the experimental situation for each participant accurately and embed self-report scales in the user interface. Moreover, VEs are able to transcend the spatial and temporal boundaries of the physical world, which further expands the possibilities for the researcher.

1.3   Social Influence Techniques in Virtual Environments

Eastwick and Gardner [7] used unstructured online virtual world There.com to test if the FITD and DITF techniques are efficient also in computer-mediated, virtual settings. It is a bridge between the previously described experiments in natural conditions and laboratory tests using VE. The experimental procedure, therefore, takes place in a VE, but the participants did not submit to it in the laboratory. Instead, they were virtual passersby, devoted to their usual activities. The experimenter, acting as an ordinary user, approached avatars standing alone and started a conversation with them, saying that he was doing a photo scavenger hunt. In the control condition, he then asked to teleport with him to another location and for permission to take a screenshot of the user's avatar on a virtual beach. In the FITD condition, the moderate request was preceded by an easy-to-meet request: "Can I take a screenshot of you:)?" and in the DITF condition, a difficult-to-meet request, regarding teleporting and taking screenshots in 50 different places, which was supposed to take two hours. Both techniques proved effective. Significantly more participants agreed to a moderate request after agreeing to a small request or refusing a large request. Additionally, the influence of the experimenter's avatar race in the DITF condition was noted. More people agreed to the request of a light-skinned avatar.

Fig. 1. Participants in virtual environments created in Garry's Mod (Half Life engine; left) and Minecraft (right)

Pochwatko et al. [16] conducted a similar study in laboratory conditions, which allowed for the continuous registration of the physiological arousal operationalized as an increase in galvanic skin response (GSR). They used a specially designed virtual environment, developed using the Half-Life game engine[1] (see Fig. 1 left). The participants were convinced that in the VE there are other people who perform miscellaneous tasks (searching for and collecting objects of various kinds, taking pictures of other users in marked places or marking boards with their initials). The draw at the arrival to the lab was manipulated in such a way that the participant was given the last of the above-mentioned tasks. Assigning participants a facade task was important as it put them in similar conditions to virtual passersby from previous studies. In order to fulfil the experimenter's request, they had to stop performing their task, as did the participants in the study of Estwick and Gardner [7]. Requests were formulated by the avatar of the experimenter's assistant or virtual agent. They concerned taking photos at the meeting point (FITD - easy request), several characteristic places on the map, which required a lot of time and travelling (DITF - difficult request) and taking a photo in one remote location (moderate request and control condition). The classic FITD and DITF effects were not replicated, but other interesting effects

---

[1] GARRYS MOD. Facepunch Studios, PC, 2006; HALF-LIFE 2. Valve Corporation. PC, 2004.

were obtained. A different pattern of physiological arousal was observed in the experimental and control groups. In the control group, participants who were less aroused complied, while in the experimental groups, those who were more aroused after the first request did so. In addition, participants agreed to fulfil requests formulated by an experimenter's assistant avatar more often than by a virtual agent. It is also associated with less agent-induced physiological arousal after formulating the request. As previous attempts to replicate social influence effects in VE have yielded mixed results, we aimed at testing whether they would be effective in certain type of VE, which offers high ecological validity and control of the confounding variables. Thus, we chose a popular multiplayer game, for the experimental scene.

2   Current Studies: Foot-in-the-face (FITF) in Laboratory Virtual Environment

Participants. 259 people participated in the study (170 in the replication of the classic procedure with and without time delay between requests and 89 in conditions with induction of guilt). The research was conducted anonymously; participants were randomly assigned experimental conditions. The experimenter's confederates were unaware of the hypotheses or the condition in which they were involved.

  Procedure. In our current studies, we used a VE created in the online version of Minecraft to deal with the limitations caused by the need to create artificial facade tasks as in the environments previously created in the Half-Life game engine. In Minecraft, the tasks for players are natural and engaging (which is confirmed, for example, by the incredible popularity of the game[2]). Therefore, there is no need to impose facade tasks that are to simulate the situation of interrupting current activities by the requester. A "survival" mode was used in which participants have to build a simple shelter and gather as many resources as possible to survive in the virtual world (in the case of an experimental situation, the test lasted three days and three nights of game time - about 40 min)[3]. Tests have shown that this task is feasible even for people who do not play computer games. At the same time, unlike the scenarios created in previous studies, the task is natural and so unstructured that it should not overburden the participants. The general scheme of the study was similar to that of Pochwatko et al. [16]. The requests, constituting the experimental manipulation, were adapted to the specificity of the environment. They involved taking photos and video clips in different places and during various activities. Participants were informed that the study was running on a public server (to keep control of the situation, this was, in fact, an isolated server in the local network). Numerous modifications were applied to simulate the presence of many players (false information about number of players when logging in, false messages about the actions of players,

---

[2] About 40 million online players around the world at the time the study was conducted.

[3] Some influence techniques (e.g. FITD) benefit from a delay between requests.

NPCs, etc.). The experimenter's confederate pretended to be a vlogger shoot-ing the footage for his YouTube channel (very popular activity for Minecraft online community). The final request concerned consent to make a clip in which the participant performs simple activities (cuts trees, builds a shelter or digs a tunnel).

As previously shown, FITD and DITF procedures are effective in VEs and lead to compliance. Regardless of whether the participant agrees or refuses to comply with the first request, the probability of consenting to the second request increases. The conclusions from the previous research were used to plan the FITF studies. The procedure used in FITD and DITF proved successful and was only slightly modified, in line with the FITF assumptions - the request sequence was changed: instead of the initial small or very large request, a request of the same degree of difficulty as the target request was used. As in the original study by Doliński [5], previously tested requests were used, which under control conditions are fulfilled by approximately half of the participants.

Additionally, for half of the participants, requests were formulated on the first and third day of game time, which was to promote the so-called "agreement route" or the FITD mechanism. In the case of the second half of the respondents, the requests were not separated in time, which was used to test the "refusal route", i.e. the DITF mechanism. In another condition, an additional variable was introduced - inducing a sense of guilt if the first request was not fulfilled (see [8]). In addition to the participants' behaviour, physiological responses were also constantly monitored. The purpose of using these measures was to verify the presence of physiological arousal at different stages of the procedure. Unpleas-ant arousal, or rather the desire to reduce it, is often mentioned as one of the mechanisms explaining the effectiveness of social influence techniques.

Apparatus. The virtual environment was presented using an Intel Core i7 2.3GHz PC with Nvidia GeForce GTX660, and BenQ short throw stereoscopic projector. GSR was recorded with a Biopac MP150 with an EDA100C amplifier.

Results. Under classic conditions (as in Dolinski 2011 [5], without inducing a sense of guilt), a number of differences in response to requests under specific conditions were observed. As assumed in the control condition, almost exactly half of the respondents agreed to the first request (49%). Both in the condition with postponed and immediate requests, significantly more people agreed to the second request after consenting to the first (agreement route), 85% and 86% respectively: postponed request $\chi^2 = 9.74$ p <.001, immediate request $\chi^2 = 8.33$ p <. 01. Significant differences were also noted on the refusal route (80% and 83% refusal) respectively: postponed request $\chi^2 = 5.69$ p <. 01, immediate request $\chi^2 = 7.59$ p <. 01 (Fig. 2). Contrary to expectations, if the first request was refused, significantly fewer people agreed to the second request. The classic FITF effect was therefore not observed. After joint analysis of the agreement and refusal routes, there are no differences between the number of participants agreeing to the target request in the experimental and control conditions (χ^2 n.s.). Therefore, the interpretation of physiological responses is limited. Nevertheless, different arousal patterns were observed in the agreement and refusal routes. Among the participants that agreed to the first request, a significant increase in physiological

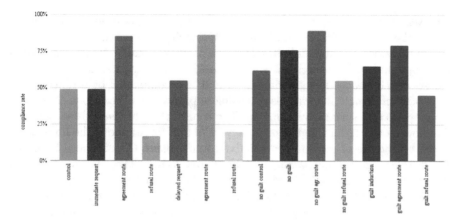

Fig. 2. Compliance rates in FITF experiments

arousal (GSR) was observed immediately after the request was formulated. The arousal decreased faster among those who agreed to the second request. On the refusal route, a significant increase in arousal after the first request was observed only in people consenting to the second request. This result is consistent with previous postulates regarding the presence of unpleasant arousal, which leads to compliance, as we observe the "arousal-consent" pattern. In the guilt induction condition, a compliance effect was observed in the guilt-free agreement route ($\chi^2 = 3.99$ p <0.05). No significant effects were observed in the remaining conditions. While there was a tendency to increase compliance in the overall results, the difference did not reach statistical significance. Similar to the previously obtained results, physiological responses in the no-guilt condition indicate an increase in arousal after the first request.

Limitations. The occurrence of certain factors limiting the scope of interpretation of the above results cannot be ruled out. First, unlike classical research, participants had to agree to come to the laboratory and use bulky equipment. Thus, The altered self perception effect could occur, i.e. perceiving oneself as a helpful person, which could have influenced their decisions. On the other hand, it may have led to the belief that they have already agreed to many things and do not need to agree to further requests, which would reduce the tendency to help. Another limitation may be the game's survival mode. On the one hand, it provides an open and ecologically valid task, but on the other hand, it can over-draw attention of participants and make them refuse more frequently.

3 Conclusions and Further Directions

The abovementioned successful replications of the sequential request techniques effect and partial replication of the FITF effect prove the possibility of conducting research on social influence in VEs and in laboratory conditions, which allows for continuous measurements of physiological responses and inference about the effectiveness of the studied techniques on the basis of more objective indicators than in the case of field studies. Still, the specificity of these influences requires further research. The dynamic development of virtual reality technology allows us to hope that such research will soon be possible on a larger scale. The number of users of VR headsets is growing rapidly. It is accompanied by an increase in the quality of devices for wireless recording of physiological reactions. There are also mass virtual social environments. The combination of these three elements will allow us to conduct high-ecologically valid and forget about the limitations caused by the use of environments derived from games. Classic social influence effects are present both in games and VEs, as well as in real life. The use of games and VEs to study them offers the possibility of better control of the experimental situation but also brings limitations that we must be aware of as researchers. Sometimes a game is just a game, but sometimes it is becoming serious.

Acknowledgments. This project has received funding from the National Science Center OPUS no. 2012/05/B/HS6/03630 We would like to thank our participants, volunteers from VRLab IPPAS panel, and our confederates: Martyna Radzikowska and Andrzej Wiatrow for countless hours in a serious game.

References

1. Blascovich, J.: A theoretical model of social influence for increasing the utility of collaborative virtual environments. In: Proceedings of the 4th International Conference on Collaborative Virtual Environments, pp. 25–30 (2002)
2. Blascovich, J., Loomis, J., Beall, A.C., Swinth, K.R., Hoyt, C.L., Bailenson, J.N.: Immersive virtual environment technology as a methodological tool for social psychology. Psychol. Inq. **13**(2), 103–124 (2002)
3. Cialdini, R.B., Vincent, J.E., Lewis, S.K., Catalan, J., Wheeler, D., Darby, B.L.: Reciprocal concessions procedure for inducing compliance: the door-in-the-face technique. J. Personal. Soc. Psychol. **31**(2), 206 (1975)
4. Daft, R.L., Lengel, R.H.: Information richness. a new approach to managerial behavior and organization design. Technical report, Texas A and M University College Station Coll of Business Administration (1983)
5. Dolinski, D.: A rock or a hard place: the foot-in-the-face technique for inducing compliance without pressure 1. J. Appl. Soc. Psychol. **41**(6), 1514–1537 (2011)
6. Dolinski, D.: Techniques of Social Influence: The Psychology of Gaining Compliance. Taylor & Francis, Milton Park (2015)
7. Eastwick, P.W., Gardner, W.L.: Is it a game? Evidence for social influence in the virtual world. Soc. Influ. **4**(1), 18–32 (2009)
8. Fourie, M.M., Rauch, H.G., Morgan, B.E., Ellis, G.F., Jordaan, E.R., Thomas, K.G.: Guilt and pride are heartfelt, but not equally so. Psychophysiology **48**(7), 888–899 (2011)
9. Freedman, J.L., Fraser, S.C.: Compliance without pressure: the foot-in-the-door technique. J. Personal. Soc. Psychol. **4**(2), 195 (1966)
10. Guadagno, R.E., Cialdini, R.B.: Persuade him by email, but see her in person: online persuasion revisited. Comput. Human Behav. **23**(2), 999–1015 (2007)
11. Hoffmann, L., Krämer, N.C., Lam-chi, A., Kopp, S.: Media equation revisited: do users show polite reactions towards an embodied agent? In: Ruttkay, Z., Kipp,

M., Nijholt, A., Vilhjálmsson, H.H. (eds.) IVA 2009. LNCS (LNAI), vol. 5773, pp. 159–165. Springer, Heidelberg (2009). https://doi.org/10.1007/978-3-642-04380-2_19

12. Kyrlitsias, C., Michael-Grigoriou, D., Banakou, D., Christofi, M.: Social conformity in immersive virtual environments: the impact of agents' gaze behavior. Front. Psychol. **11**, 2254 (2020)

13. Nass, C., Moon, Y., Carney, P.: Are people polite to computers? Responses to computer-based interviewing systems 1. J. Appl. Soc. Psychol. **29**(5), 1093–1109 (1999)

14. Nass, C., Steuer, J., Tauber, E.R.: Computers are social actors. In: Proceedings of the SIGCHI Conference on Human Factors in Computing Systems, pp. 72–78 (1994)

15. Park, S., Catrambone, R.: Social facilitation effects of virtual humans. Human Factors **49**(6), 1054–1060 (2007)

16. Pochwatko, G., Oseka, L., Świdrak, J.: Wplyw spoleczny w realnej i wirtualnej rzeczywistosci–uleglosc w relacji z awatarami, agentami i robotami. Nowoczesne technologie XXI w.–przeglad, trendy i badania. Tom p. 274 (2019)

17. Reeves, B., Nass, C.: The Media Equation: How People Treat Computers, Television, and New Media Like Real People. Cambridge University Press, Cambridge (1996)

18. Zanbaka, C.A., Ulinski, A.C., Goolkasian, P., Hodges, L.F.: Social responses to virtual humans: implications for future interface design. In: Proceedings of the SIGCHI Conference on Human Factors in Computing Systems, pp. 1561–1570 (2007)

Participatory Action for Citizens' Engagement to Develop a Pro-environmental Research Application

Anna Jaskulska[1,3], Kinga Skorupska[1,2(✉)], Zuzanna Bubrowska[1],
Kinga Kwiatkowska[1], Wiktor Stawski[1], Maciej Krzywicki[1],
Monika Kornacka[2], and Wiesław Kopeć[1,2,3]

[1] Polish-Japanese Academy of Information Technology, Warsaw, Poland
kinga.skorupska@pja.edu.pl
[2] SWPS University of Social Sciences and Humanities, Warsaw, Poland
[3] Kobo Association, Warsaw, Poland

Abstract. To understand and begin to address the challenge of air pollution in Europe we conducted participatory research, art and design activities with the residents of one of the areas most affected by smog in Poland. The participatory research events, described in detail in this article, centered around the theme of ecology and served to design an application that would allow us to conduct field research on pro-environmental behaviours at a larger scale. As a result we developed a research application, rooted in local culture and history and place attachment, which makes use of gamification techniques. The application gathers air quality data from the densest network of air pollution sensors in Europe, thereby aligning the visible signs of pollution in the app with the local sensor data. At the same time it reinforces the users' pro-environmental habits and exposes them to educational messages about air quality and the environment. The data gathered with this application will validate the efficacy of this kind of an intervention in addressing residents' smog-causing behaviours.

Keywords: Participatory design · User research · Citizen engagement · Citizen science · Application design · Place attachment

1 Introduction and Related Works

Addressing the challenge of air pollution is a pressing matter for European governments, as despite increasingly better policies and green energy solutions Europe falls short of its zero pollution goal. In many places air pollution levels are above the alert thresholds set by WHO, which results in increased prevalence of health issues and higher mortality [1]. In Poland, household heating is the source of almost half of PM10 and PM2.5 and 90% of PAH emissions [4].

C. Biele et al. (Eds.): MIDI 2021, LNNS 440, pp. 198–207, 2022.
https://doi.org/10.1007/978-3-031-11432-8_20

To better understand this problem and come up with potential bottom-up solutions we conducted an extensive literature review and engaged in field research with residents of the area most affected by smog in Poland. The town of Myszków, is located in the Silesian Voivodeship, in Upper Warta River Depression. As a result of this location, smog generated by neighboring towns and villages settles in Myszków, which qualifies the town as one of the most polluted towns in Poland [2,3]. The local authorities are aware of the problem of smog and have previously installed a couple of smog sensors, but despite such programs and efforts to subsidize clean energy solutions the problem persists. This awareness facilitated the town's active participation in the VAPE ecological project conducted, among others, by the XR Lab of the Polish-Japanese Academy of Information Technology (PJAIT), the Institute of Psychology of the Polish Academy of Sciences (IP PAN), Warsaw University Faculty of Economic Sciences and Norwegian Institute For Air Research (NILU). The main goal of the VAPE project was to investigate how multi-sensory virtual experiences affect people's environmental behaviors. In this context we organized multiple participatory research, art and design activities with citizens. These were centered around the theme of ecology and aimed to co-develop an application that would allow us to conduct field research on how to encourage pro-environmental behaviours at a larger scale.

Based on our participatory design [8,9,11], citizen science [14] and living lab [10] experience, and an extensive literature review, especially in the area of environmental psychology and phenomenological geography as well as sociology we have developed an approach useful to engage citizens in pro-environmental projects by rooting these projects in the local environment, history and culture. This approach is backed by theories of place attachment [13] and place identity. Identification with a certain places can result in an increased quality of life and a greater commitment to protect one's habitat, neighbourhood and public spaces. This, in turn helps develop positive environmental behaviours [16]. We also made use of the concept of cognitive dissonance, which can be a powerful motivating force [6]. We made use of this concept together with the ideas of both the "foot in the door" technique [7], and "nudges" [15] as they can be helpful in shaping ecological behaviours. We were also inspired the role of children and young adults as environmental educators and ambassadors among their families [5,12].

2 Methods and Results

2.1 Participatory Research and Citizen Engagement

Our activities with residents of Myszków were divided into three stages:

1. Participatory workshops at the local vineyard;
2. In-depth ethnographic research including digital ethnography and in-depth interviews with residents in their everyday contexts, city walks (both in the field and with the help of evaluative maps and questionnaires on multisensory perception of places important to the residents of Myszków);
3. Eco-picnic in the local activity park.

Together with the residents we concentrated on the awareness of environmental issues and social causes of non-ecological behavior. We also focused on issues related to the residents' attachment to their city to find ways to engage them in pro-environmental activities. All of these events allowed the residents to open up to the issues raised by the VAPE project and engage in its activities.

2.2 Participants

Almost a hundred citizens of Myszków of all ages took part in our activities and research. Some of them were activists, involved in local formal and informal organizations and interest groups while others were just interested in what was happening in their city. Overall, we identified four types of residents in Myszków in terms of their attitudes towards the environment and air pollution:

Environmentally conscious with resources	Residents who are environmentally conscious and care about the environment and have sufficient financial resources and opportunities (e.g. local entrepreneurs engaging in actions such as forest cleanups).
Environmentally conscious with low resources	Residents who try hard to care for the environment despite logistic and financial difficulties (e.g. living at the outskirts of the city, without access to all utilities, like the connection to the gas installation).
Non-eco-friendly with resources	Residents who do not care for the environment despite having sufficient resources and possibilities, due to unawareness and ignorance of ecological issues.
Non-eco-friendly with low resources	Residents who do not care for the environment, who also lack the financial means and cannot afford more sustainable practices.

Fig. 1. The outdoor locations: the vineyard and the activity park during project events

2.3   Locations

Our participatory research activities with citizens took place directly in the field (see Fig. 1) which allowed us to eliminate the laboratory environment and workshop rooms. To organize our events we selected the only two "places of interest" in Myszków listed on TripAdvisor, a popular travel platform: a local vineyard and an activity park. Both provided outdoor facilities, which was important due to the COVID-19 pandemic, and were well-aligned with the green theme of our project. What follows is a detailed description of the scenarios of these activities as well as a summary of insights gained in their course.

2.4   Participatory Workshops at the Local Vineyard

The participatory processes at the vineyard were supported by activities related to air and water pollution, sustainability and environmental education as well as presentations of new technologies that can support these. Participants took part in a guided tour of the vineyard and an exhibition of macro-photographs of insects threatened with extinction by climate change (see Fig. 2).

Fig. 2. Guided tour of the vineyard and macro photography exhibition

At the vineyard we also placed two artworks by Aleksandra Karpowicz, a London-based video artist. One of them, called "Lungs for sale", which depicts the coughing artist, was played on a loop on a smartphone buried in the ash from a campfire. The visitors, tired of coughing, could walk into a row of vines where, in the second video, Alexandra is dancing in the meadow, and her message "wake up" can be heard in the background (see Fig. 3).

Fig. 3. "Lungs for sale" and "Wake up", video artworks by Aleksandra Karpowicz

Aleksandra Karpowicz also prepared a video manifesto in which she addressed issues related to climate change, air and earth pollution, showing them from the social, political and economic points of view. We built a wooden TV set stand in order to fit it into the natural surroundings where the video was played. These works of art served the same purpose as a lecture would: they introduced the participants to the topic of the workshop. We used them to provoke discussions with various groups of residents, who were more open to share their thoughts because the artist did it first. The second empowerment element consisted of demonstrations of solutions related to virtual, augmented and mixed reality. During our workshop with the participants we brainstormed, facilitated discussions and collected insights.

2.5   In-depth Interviews and Ethnographic Research

In order to deepen the insights gathered during the participatory workshops, we conducted qualitative research including individual and group in-depth interviews; digital ethnography - comprehensive research of local websites and social networks, like public and private groups for residents, profiles of the town, municipal police, social and cultural organizations, archives of town history, etc.; ethnographic walks (including a "virtual one" using only the map) combined with multidimensional descriptions of selected places; exploratory walks of people who have never been to Myszków - using urban signposting and data from geolocation-based games and travel applications (see Fig. 4). Additionally, a series of quantitative studies in the form of online surveys are underway.

Fig. 4. Examples of Digital Ethnography research from Myszków's social media and cards from city walks with photos printed with a portable photo printer

2.6   Family Ecological Picnic

During the eco-picnic together with the inhabitants of Myszków we conducted activities directly involving residents. Adults could participate in the creative process to paint Myszków as the city of their dreams and children played with colouring books on waste segregation (see Fig. 5).

We gave the citizens of Myszków the possibility to test our project solutions in VR, AR and mobile apps. We also showcased 3D printers used in the VAPE

Fig. 5. Artistic activities with residents of all ages during the eco-picnic

project to prepare objects for upcoming research on the perception of touch in virtual reality (see Fig. 6). We invited a talk on violent weather phenomena and organized sports activities, related to the seasons, such as beach volleyball or curling (presented as a board game).

Fig. 6. VR, AR & 3D printing during eco-picnic

3   Discussion

In the course of our research we identified several key areas of non-ecological behavior among residents that contribute to smog formation. These were mostly related to heating their homes and included lack of access to municipal heating, no access to gas infrastructure in the neighbourhood, and thus gas heating, the use of old stoves without adequate filters, the use of improper techniques of burning in the stove (increasing the amount of smoke) and burning garbage. Some of them are determined by administrative reasons, for example, receiving EU subsidies to replace the stove with a more ecological one is possible only after the specific house has been modernized using the interested person's own funds. In some cases the problem is caused by poverty and lack of funds to buy appropriate heating fuel. We have excluded these factors from our area of focus as they are related to the environmental and social policies, and can not change as a result of our project. The city authorities are aware of these problems and try to actively counteract them, e.g. by creating their own programs for stove replacement without additional conditions.

Another non-ecological activity in Myszków is using garbage to heat homes: household litter, used tires or old painted furniture. The scale of the problem is illustrated by the reports from the municipal police interventions. There are even cases of burning garbage in new stoves designed for ecological fuel. The problem persists even though waste collection fees do not depend on quantity, so throwing garbage into the stove does not result in direct savings. Causes of litter burning, according to our participatory research include: lack of knowledge about the consequences of burning garbage, viewing it as a way to get rid of garbage as quickly as possible, e.g. burning leaves without waiting for them to be collected, lack of knowledge about the possibility to dispose of difficult garbage such as used tires, old furniture, renovation waste, free of charge at the local selective waste collection point, viewing garbage as valuable fuel, especially in the cooler seasons of autumn and spring. In the course of further analysis we decided that trying to influence this behavior through instilling proper garbage disposal and sorting, as well as choosing the right fuel for heating, ought to become a key issue addressed in our mobile field research application.

4   Mobile Field Research Application

The Vape Mobile app combines a Match-3 game concept (inspired by Bejeweled, Candy Crush) with a simulation of how the air quality in Myszków affects both the town and the daily life of its inhabitants. Based on the "foot in the door" technique, the garbage sorting and choosing appropriate heating sources in the game is meant to be the first step to feeling environmentally engaged and making pro-environmental choices in the real world.

Fig. 7. Screens from the pro-environmental research application to address the problem of smog. From left to right: 1) The user sees that the local air quality is not very good 2) they can engage with the game where they sort trash, or choose better energy sources 3) their performance is saved 4) if air quality improves, then the application skin changes

The application was designed to have simple mechanics and be fit for people of different ages, including children and older adults. The application was to be set in the local context - the player improves their eco-Myszków and progress is achieved by sorting garbage and choosing appropriate fuels for heating (see Fig. 7). Gamification elements such as educational quizzes, quests or leaderboards ought to encourage regular use without being time-consuming. The game also ought to provide exposure to smog visualizations to continue laboratory studies in the field. By design the participant, in order to gain additional points in the game, in the frequency determined by the researchers, is exposed to visualizations of pollution or clean air. The user can also compare the state of 'their Myszków' and the real life one, based on the actual readings from air pollution sensors (this functionality uses the sensor network in Myszków).

The app was created with the use of Unity for Android and common algorithms used in Match-3 type games.

All assets used in the game are either free to use or were created by us using Procreate or Clip Studio Paint. During our research we collected references to properly portray Myszków and make sure it is recognizable in the app at first glance. Many landmarks such as the historical railway station, a local palace and the bandshell from activity park were used to achieve this. The information on recycling trash is based on regulations valid in Myszków.

5   Conclusions and Further Work

Thanks to the participatory research and art activities we were able to create a field research application for raising awareness about smog. Based on our experience, to engage in similar project activities, we can recommend working closely with local authorities, institutions, community organizations and activists as well as eco-friendly stakeholders. Digital ethnography, including a review of local social networks, is also worth including in this type of research. In the next steps we will continue to develop and test new application functionalities. After completing the first cycle of quantitative remote survey-based research we will be able to inform the application design with insights from the results of survey data analyses. Overall, empowerment and community inclusion in the research, design and development processes, for which both art and new technologies can be used are powerful vessels to discover real needs of potential users. Such activities ought to constitute a crucial part of impact-focused social change projects enabled by novel technologies.

Acknowledgments. The VAPE project, which made this research possible, received funding from the IdeaLab competition for interdisciplinary research projects. It was funded by the European Economic Area Financial Mechanism 2014–2021 (grant number 2019/35/J/HS6/03166). We would also like to thank the many people and institutions gathered together by the distributed Living Lab Kobo and HASE Research Group (Human Aspects in Science and Engineering) for their support of this study.

References

1. New WHO global air quality guidelines aim to save millions of lives from air pollution. https://www.who.int/news/item/22-09-2021-new-who-global-air-quality-guidelines-aim-to-save-millions-of-lives-from-air-pollution
2. Smogowi liderzy - ranking polskich miast z najbardziej zanieczyszczonym powietrzem, October 2019. https://smoglab.pl/smogowi-liderzy-ranking-miast/
3. Najbardziej rakotwórcze powietrze w polsce, November 2020. https://polskialarmsmogowy.pl/2020/11/najbardziej-rakotworcze-powietrze-w-polsce-pas-prezentuje-ranking-miast/
4. Poland's informative inventory report, March 2021. https://cdr.eionet.europa.eu/pl/eu/nec_revised/iir/envyei5sq/
5. Blanchet-Cohen, N., Reilly, R.: Immigrant children promoting environmental care: enhancing learning, agency and integration through culturally-responsive environmental education. Environ. Educ. Res. **23**, 553–572 (2017)
6. Festinger, L.: A Theory of Cognitive Dissonance. Stanford University Press, Stanford (1962)
7. Freedman, J.L., Fraser, S.C.: Compliance without pressure: the foot-in-the-door technique. J. Pers. Soc. Psychol. **4**(2), 195–202 (1966)
8. Kopeć, W., et al.: Participatory design landscape for the human-machine collaboration, interaction and automation at the frontiers of HCI (PDL 2021). In: Ardito, C., et al. (eds.) INTERACT 2021. LNCS, vol. 12936, pp. 564–569. Springer, Cham (2021). https://doi.org/10.1007/978-3-030-85607-6_78
9. Kopeć, W., Nielek, R., Wierzbicki, A.: Guidelines towards better participation of older adults in software development processes using a new spiral method and participatory approach. In: Proceedings of the 11th International Workshop on Cooperative and Human Aspects of Software Engineering, CHASE 2018, pp. 49–56. Association for Computing Machinery, New York (2018). https://doi.org/10.1145/3195836.3195840
10. Kopeć, W., Skorupska, K., Jaskulska, A., Abramczuk, K., Nielek, R., Wierzbicki, A.: LivingLab PJAIT: towards better urban participation of seniors. In: Proceedings of the International Conference on Web Intelligence, WI 2017, pp. 1085–1092. ACM, New York (2017). https://doi.org/10.1145/3106426.3109040
11. Kopeć, W., et al.: VR with older adults: participatory design of a virtual ATM training simulation. IFAC-PapersOnLine **52**(19), 277–281 (2019). https://doi.org/10.1016/j.ifacol.2019.12.110. https://www.sciencedirect.com/science/article/pii/S2405896319319457. 14th IFAC Symposium on Analysis, Design, and Evaluation of Human Machine Systems, HMS 2019
12. Leppänen, J., Haahla, A., Lensu, A., Kuitunen, M.: Parent-child similarity in environmental attitudes: a pairwise comparison. J. Environ. Educ. **43**, 162–176 (2012). https://doi.org/10.1080/00958964.2011.634449
13. Lewicka, M.: Place attachment: how far have we come in the last 40 years? J. Environ. Psychol. **31**(3), 207–230 (2011). https://doi.org/10.1016/j.jenvp.2010.10.001
14. Skorupska, K., Jaskulska, A., Masłyk, R., Paluch, J., Nielek, R., Kopeć, W.: Older adults' motivation and engagement with diverse crowdsourcing citizen science tasks. In: Ardito, C., et al. (eds.) INTERACT 2021. LNCS, vol. 12933, pp. 93–103. Springer, Cham (2021). https://doi.org/10.1007/978-3-030-85616-8_7
15. Thaler, R.H., Sunstein, C.R.: Nudge. Yale University Press, New Haven, CT and London (2008)
16. Uzzell, D., Moser, G.: Environment and quality of life. Eur. Rev. Appl. Psychol. **56**(1), 1–4 (2006). https://doi.org/10.1016/j.erap.2005.02.007

Use of Virtual Reality in Psychology

Arhum Hakim[1](✉) and Sadaf Hammad[2]

[1] School of Social Sciences and Humanities, National University of Sciences and Technology (NUST), Islamabad, Pakistan
arhumhakim@gmail.com
[2] Video Surveillance Lab PNEC-NUST, Karachi, Pakistan

Abstract. The field of psychology is advancing with incorporation of immersive technologies and Virtual Environments (VE) in research, treatment, assessment and learning etc. Virtual Reality (VR) is one of the focused sub-domains of immersive technologies that is being used vastly in psychology. This also comes under the paradigm of cyberpsychology. The advancements reported in this area bring the need to understand the benefits provided by virtual reality with respect to human interaction and behavior etc. The published literature available discusses the development and evolution of cyber psychology specifically with respect to virtual reality. To the best of our knowledge the latest developments due to virtual reality in various sub fields of psychology have not been collectively documented so far. This paper provides highlights of the developments and transformations linked with the usage of virtual reality with recommendations about future studies and provides the reader a broader and informed view of this cross-disciplinary area. Moreover, the paper also provides limitations that exist, and considerations required while using VR for a specific treatment or experimentation.

Keywords: Immersive technology · Virtual Reality · Cyberpsychology · Psychology

1 Introduction

Psychology is a vast discipline involving scientific investigation of mind and behavior. It includes the understanding of mind and how it affects a person's behavior. Overtime, the field has been expanded into multiple sub-domains including clinical, experimental, educational, child, rehabilitation, sports, and social psychology etc. Similarly, the field of psychology consists of multiple professionals who mainly assess and treat mental health issues, counselors who counsel individuals, and professionals who work for the betterment of a team/organization. Psychology as a field has further progressed with the application of immersive technologies and Virtual Environments (VE). Virtual Reality (VR) is one of the focused sub-domains of immersive technologies that is being used vastly in psychology. In VR the user operates through three dimensional VE and interacts with the environment using sensory inputs from especially designed head and hand gears. The VR headsets are wearable devices having display optic for each eye leveraging 360° view and have onboard or off-board/wired processing options. VR usage in the field of

C. Biele et al. (Eds.): MIDI 2021, LNNS 440, pp. 208–217, 2022.
https://doi.org/10.1007/978-3-031-11432-8_21

psychology has enhanced not only the interaction between people but has also facilitated the researchers to conduct studies that are impossible to carry out in the real world. This research paper highlights the usage of VR and related advancements in major sub-fields of psychology.

The rest of the paper is organized as follows. Section 2 briefly represents the background research reported in the area. Applications of VR in various psychology sub-domains are presented in Sect. 3. Section 4 highlights some limitations. Section 5 provides the discussion with some future research dimensions and Sect. 6 the conclusion.

2   Background

The advancements in the field of psychology in relation to technology has been discussed in literature by the research community. Findings reveal that, in current times, virtual reality appear to be one of the beneficial tools for obtaining effective outcomes with respect to therapy and rehabilitation among patients [1]. Different studies have also highlighted the developments in specific areas of psychology like assessment, diagnosis, treatment, research, and training of trainees etc. [2, 3]. Keeping in view the structure of the paper, we have discussed the related applications in Sect. 3. To the best of our knowledge, there is still a requirement of a document providing a broader spectrum of current developments in psychology with respect to VR. Such a document can facilitate the readers and researchers to acquire a broader picture and significance of incorporating the emerging technologies in the field linked to humans and their mental health. In this paper, we have tried to emphasize the overall usage and benefits that VR has provided in different sub-domains of psychology. Adding to it, gaps in previous studies with recommendations have also been discussed.

3   Application of VR in Major Sub-domains of Psychology

3.1   Clinical Psychology

3.1.1   Application in Therapy

Majority of the researchers have documented developments specifically in the domain of mental health. Effective and progressive work regarding VR is being done in treating various mental health issues. The overall treatment aided by VR involves traditional treatment coupled with VR exposure based on the individual needs of the subject. The individualized aspect is determined by thorough and detailed assessment of the individual's concern, its intensity and overall condition [2]. The effectiveness of VR in terms of satisfaction of individuals has also been assessed through a survey conducted on US Soldiers. The findings of the survey revealed that most of the individuals prefer technology-aided therapy for mental health as compared to the traditional ones [20]. Brief summary of the applications of VR in therapy are appended below:

Height Phobia. Literature indicates effectiveness of virtual environments for exposure therapy to treat phobias. An experimental study conducted by Emmelkamp *et al.* [5] explored the efficacy of utilizing virtual environments coupled with exposure therapy for treating height phobia. Findings reveal that VR based exposure therapy is very effective compared to traditional exposure therapy as it corroborates the safety of client and doesn't lead to fear responses that are generally experienced and high in intensity in non-virtual scenarios. It also enables the therapist to facilitate clients with weak imagination. Further, it is found to be cost effective and convenient [3]. Similar experiments and analysis was also indicated by other researchers. For example, in a recent Korean society based systematic review, effectiveness of VR psychotherapy has been analyzed for fear of height with some considerations like cyber sickness and the level of presence felt during the process [4–6].

Anxiety Disorders. Clients having anxiety disorders like Social Anxiety Disorder (SAD), Panic Disorder and Generalized Anxiety Disorder have also benefitted from VR based therapy [2, 7–9]. Further, VR is also being used for alleviating anxiety symptoms through relaxation and mindfulness techniques including breathing exercises [2]. A recent study revealed that VR is also providing stress management and social support to general public during pandemic by screening 3D movie (The Secret Garden) using mobile enabled cardboard VR headsets [10].

Eating Disorders. Studies on eating disorders indicate that VR scenarios and VR based therapy bring more improvement in symptoms of patients compared to traditional therapy. In this domain, the virtual scenarios primarily focus on cravings, body image, and emotional regulation skills [11, 12]. Moreover, Exergames involving various exercises use digital gaming for treating obesity [13].

Addictions. Combined with traditional therapy, VR has been found facilitative in reducing and preventing relapse among clients having substance addiction [14–16]. It majorly focuses on cue exposure and craving of substance [17].

Pain Management. Medicines for pain management can lead to side effects, over usage and dependence in patients [18]. VR is being recently studied for pain management as well. It introduces management of pain through virtual scenarios that facilitate distraction, shifting focus and developing skills to manage pain [19].

Post Traumatic Stress Disorder (PTSD). Literature indicates favorable outcomes in case of PTSD as the symptoms experienced due to traumatic event are found to be reduced with the use of VR [9, 21, 22]. Though more extensive research is required as studies so far have included small sample size, lack of randomization and control groups [8, 23].

False Sensory Experience and Reduced Behavioral Functioning. VR is also being researched for psychotic disorders in which the client has no contact with reality and experience delusions (false beliefs) and hallucinations (false sensory experience) with negative symptoms like reduced emotional and behavioral functioning [9]. Also some related studies conducted reveal that VR with a combination of traditional therapies can help schizophrenic patients to improve their social skills and problem solving linked to daily routine [24, 25]. Further, it is also found to be effective for reducing hallucinations and delusions [26].

3.1.2 Application in Psychological Assessment

It is quite difficult to assess individual's experience, cognitions and behavior with paper and pencil tests. VR enables the professionals to assess the patient under real life situations and provides authentic understanding. This includes individual's current state based on his/her age and real life experiences thereby making the assessment as accurate as possible. Further, executive functioning of individuals can also be evaluated using VR. This relates to disorders linked with memory issues like Alzheimer's, Parkinson's, and traumatic brain injury [27]. A recent experimental study presents that VR based assessment is equally effective compared to the classical way of testing. It provides the clinician to monitor patient's physiological changes smoothly and helps make the overall assessment comprehensive and accurate [28].

3.1.3 Application in Clinical Training

VR technology is also being applied for training clinical professionals. Skill development can be improved with the help of Virtual Environment (VE) as it can replicate participants and patients virtually and can provide feedback to the trainee therapist. Similarly, system generated virtual practitioners can be offered to the participants using VR and Artificial Intelligence (AI). The virtual practitioner bot can facilitate in recording non-verbal expressions and verbal communication of participant that can further aid in communication with client [29].

3.2 Experimental Psychology

VR has facilitated the researchers in studying psychological phenomenas by simulating real life situations that include cognitive functions like attention, memory, perception, and problem solving etc. Virtual Environment here aids in formulating ecologically valid and flexible settings. It also helps generate varying stimuli and can be used to evaluate multiple responses of the subject. This in-turn enables to evaluate complex behaviors of subjects under different situations [30]. Similarly, a recent review of 2019 about episodic memory research also supports the contribution of VR [31].

3.3 Child/Developmental Psychology

Neuro developmental disorders are common in children. This involves impairment/disturbance in child's overall functioning related to social life, personal life, education life and motor behavior [32]. In case of neurodevelopmental disorders like autism, intellectual disability, and communication disorders, the focus of intervention is primarily on communication skills. VR has been used to facilitate children with autism in gaining emotional understanding and social conversation as indicated by the studies of Cheng *et al.* [33] and Horace *et al.* [34]. Some researchers have tried to assess effectiveness of VR for children with specific developmental issues though due mixed opinions, it requires further studies to evaluate its effectiveness [35].

3.4 Educational Psychology

The VR technology is also being researched for learning and educational purposes. It can facilitate teachers to make the trainee therapists and students understand complex topic [36]. Prong *et al.* reports that students studying biology lesson using Virtual Environment indicate high motivation and interest as compared to group of students who are taught using conventional multimedia slides [37]. Overall, VR aided learning boosts motivation, attention and interest [37, 38]. VR further facilitates in visualizing the concepts in a better way that leads to enhanced understanding of concepts.

3.5 Rehabilitation Psychology

Rehabilitation psychology is a process that tends to retain plasticity of human brain. The clinical diseases specifically related to brain lead to reduced plasticity. This requires regular training to support creation of new neural pathways [39]. VR is being utilized for numerous medical conditions like stroke, cerebral palsy, traumatic brain injury, and spinal cord injury etc. [40]. Some rehabilitation VR based solutions such as Timocco have been introduced that enables the user to practice rehabilitation at home [41].

3.6 Sports Psychology

VR technology is being used for training of athletes by giving feedback about their performance during practice. Literature indicates that performance of players trained in VR is comparable to players trained in real life environment [42]. A recent study by Farlet *et al.* indicates that players' sensory and motor coordination are enhanced using virtual settings [43].

3.7 Social Psychology

VR is also being used in social psychology which includes social and behavioral constructs that are quite difficult to assess in real life [44]. The social constructs include prosocial, aggressive, and discriminating behaviors etc. These social constructs/phenomenas are being studied in depth using Virtual Environments. VR based research provides full control to the observer making it ecologically valid. Behavioral tracing is one of the techniques being used to quantify and understand the construct. In a study by Rizzo *et al*, head rotation was observed with respect to the construct of attention in a virtual classroom and the symptoms of inattention and hyperactivity were assessed [45]. Further, in [46] author have used virtual settings to evaluate the influence of proxemics (distance among individuals affecting comfort level) and gender of instructor on learning of students [47].

4 Limitations

VR is quite effective for treatment and research purposes but it's usage can also impact the individuals negatively. Literature reveals some of the possible effects like sickness caused by motion (disturbance in balance), seizures caused by light, physical injuries and fatigue due to prolonged usage, migraines, nausea, trauma, decline in cognitive performance and feelings of physical discomfort [48]. In addition to negative emotions, issues like VR simulation cost, patients getting addicted to VR experience, awareness and attitude of clinicians toward VR are also noteworthy [49]. Therefore, it is crucial for the professionals and researchers to keep the adverse impacts in mind while using VR as an augmenting tool.

5   Discussion

Over the years, VR usage has expanded among different areas of psychology making it a promising and revolutionary tool. In this paper, the benefits and advancements with respect to VR in psychology from assessment, treatment and research perspective have been discussed. The literature indicates numerous advantages, for example in clinical and developmental domain, studies indicate that use of VR has effective outcomes with respect to assessment, clinician training and treatment of various mental health disorders. Similarly, VR has been very beneficial in experimental psychology. It helps better understand and study complex phenomenas in relatively efficient way compared to traditional research by making it ecologically valid. In addition, VR has transformed research in the domain of social psychology as it is practically impossible to study implicit constructs in real life situations. VR benefits have also been expanded to the field of sports. Similarly, VR is also effective for teaching and learning purposes as it enables to visualize the concepts. VR has also been instrumental in rehabilitation domain as it facilitates the patients in gaining functional ability and enhanced quality of life.

The published literature provides a pool of studies catering VR and its application in major domains of psychology but there are some gaps that can be addressed in future research. Some of the major gaps are appended below for quick reference:

- In clinical area, there is inadequate expertise of clinicians in using VR and lack of understanding about the suitability of VR for different individuals. Attitude of patients and professional regarding VR also adds to the gap [50].
- Using VR with children having developmental issues indicate mixed results [51] and the studies have mostly focused on treatment of specific disorders [52, 53].
- Mixed results have been reported for using VR after treatment and at follow up in rehabilitation [54].
- There is a need of standardization (psychometric properties) of VR based assessment method for generalizable results [55].
- Some gaps linked to experiments are indication of mixed results and their relation to factors like presence felt in virtual scenario of an experiment [28].
- For social constructs, issues in construct validity (definition and assessment of construct) and oversimplification of observations noted during research are indicated [47].
- Studies conducted on VR based learning indicate use of inadequate methods to assess learning outcomes [56].
- Few number of studies on skill acquisition of athletes have been reported involving old VR devices. So, latest VR tools need to be explored [43].

6 Conclusion

Over the years, VR technology has transformed the areas of psychology that includes assessment, training, research and treatment etc. This paper briefly represents the usage of Virtual Environment (VE) in different sub fields of psychology. Though numerous research has been published in this area, however, mostly the findings are derived from experimentation on a limited sample size and under controlled settings. Therefore, there is a need to have standardized design, methodology and data that can be used to gauge the effectiveness of VR in sub domains of psychology. Further, work on personal factors of patients, construct validation and psychometric properties of VR can enhance VR applications in the field of psychology.

References

1. Rizzo, A.A., Schultheis, M.T., Kerns, K., Mateer, C.: Analysis of assets for virtual reality applications in neuropsychology. Neuropsychol. Rehabil. **14**(1), 207–239 (2004)
2. Maples-Keller, J.L., Bunnell, B.E., Kim, S.J., Rothbaum, B.O.: The use of virtual reality technology in the treatment of anxiety and other psychiatric disorders. Harv. Rev. Psychiatry **25**(3), 103–113 (2017)
3. Martin, S.: Virtual Reality Might Be the Next Big Thing for Mental Health. Scientific American Directorate (2019). https://blogs.scientificamerican.com/observations/virtual-rea lity-might-be-the-next-big-thing-for-mental-health/ . Accessed 15 Nov 2021
4. Liu, T., Tang, Z.: Application of virtual reality technology in clinical psychology. In: International Conference on Computer Information and Big Data Applications (CIBDA) (2020)
5. Emmelkamp, P.M., Krijn, M., Hulsbosch, A.M., de Vries, S., Schuemie, M.J., Van der Mast, C.A.: Virtual reality treatment versus exposure in vivo: a comparative evaluation in acrophobia. Behav. Res. Ther. **40**(5), 509–516 (2002)
6. Moon, J.C., Jeesu, K., Yeoung-Su, L., Hyung, W.K.: Domestic trend analysis of virtual reality therapy for the treatment anxiety disorders. J. Orient. Neuropsychiatry **31**(4), 279–288 (2020)
7. Robillard, G., Bouchard, S., Dumoulin, S., Guitard, T., Klinger, E.: Using virtual humans to alleviate social anxiety: preliminary report from a comparative outcome study. Stud. Health Technol. Inform. **154**, 57–60 (2010)
8. Pitti, C., et al.: Agoraphobia: combined treatment and virtual reality. Preliminary results. . Actas espanolas de psiquiatria **36**(2), 94–101 (2008)
9. American Psychiatric Association. Diagnostic and statistical manual of mental disorders: DSM-5 (2013)
10. Riva, G., Wiederhold, B.K.: How cyberpsychology and virtual reality can help us to overcome the psychological burden of coronavirus. Cyberpsychol. Behav. Soc. Netw. **5**, 227–229 (2020)
11. Marco, J.H., Perpina, C., Botella, C.: Effectiveness of cognitive behavioral therapy supported by virtual reality in the treatment of body image in eating disorders: one-year follow-up. Psychiatry Res. **209**, 619–625 (2013)
12. Ferrer-Garcia, M., et al.: A randomized trial of virtual reality-based cue exposure second-level therapy and- cognitive behavior second-level therapy for bulimia nervosa and binge-eating disorder: outcome at six-month followup. Cyberpsychol. Behav. Soc. Netw. **22**, 60–68 (2019)
13. Lyons, E.J.: Cultivating engagement and enjoyment in exergames using feedback, challenge, and rewards. Games Health J. **4**, 12–18 (2015)

14. Choi, J.S., et al.: The effect of repeated virtual nicotine cue exposure therapy on the psychophysiological responses: a preliminary study. Psychiatry Investig. **8**(2), 155–160 (2011)
15. Riva, G., Bacchetta, M., Baruffi, M., Molinari, E.: Virtual reality–based multidimensional therapy for the treatment of body image disturbances in obesity: a controlled study. Cyberpsychol. Behav. **4**, 511–526 (2001)
16. Pericot-Valverde, I., Secades-Villa, R., Gutiérrez-Maldonado, J.: A randomized clinical trial of cue exposure treatment through virtual reality for smoking cessation. J. Subst. Abuse Treat. **96**, 26–32 (2019)
17. Segawa, T., et al.: Virtual reality (VR) in assessment and treatment of addictive disorders: a systematic review. Front. Neurosci. **13**, 1409 (2019)
18. Wiederhold, B.K., Soomro, A., Riva, G., Wiederhold, M.D.: Future directions: advances and implications of virtual environments designed for pain management. Cyberpsychol. Behav. Soc. Netw **17**, 414–422 (2014)
19. Chan, E., Foster, S., Sambell, R., Leong, P.: Clinical efficacy of virtual reality for acute procedural pain management: a systematic review and meta-analysis. PLoS ONE **13**(7), e0200987 (2018)
20. Wilson, J., Onorati, K., Mishkind, M., Reger, M., Gahm, G.A.: Soldier attitudes about technology-based approaches to mental healthcare. Cyberpsychol. Behav. **11**, 767–769 (2008)
21. Opriş, D., Pintea, S., García-Palacios, A., Botella, C., Szamosközi, Ş, David, D.: Virtual reality exposure therapy in anxiety disorders: a quantitative meta-analysis. Depress Anxiety **29**, 85–93 (2012)
22. Difede, J., Hoffman, H.G.: Virtual reality exposure therapy for world trade center post-traumatic stress disorder: a case report. Cyberpsychol. Behav. **5**, 529–535 (2002)
23. Rizzo, A.S., et al.: Development and early evaluation of the virtual Iraq/Afghanistan exposure therapy system for combat-related PTSD. Ann. NY. Acad. Sci. **1208**, 114–125 (2010)
24. Rus-Calafell, M., Gutiérrez-Maldonado, J., Ribas-Sabaté, J.: A virtual reality-integrated program for improving social skills in patients with schizophrenia: a pilot study. J. Behav. Ther. Exp. Psychiatry **45**, 81–89 (2014)
25. Fernández-Sotosa, P., Fernández-Caballero, A., Rodriguez-Jimenez, R.: Virtual reality for psychosocial remediation in schizophrenia: a systematic review. Eur. J. Psychiatry **34**, 1–10 (2020)
26. Veling, W., Moritz, S., Van Der Gaag, M.: Brave new worlds—review and update on virtual reality assessment and treatment in psychosis. Schizophr. Bul. **40**(6), 1194–1197 (2014)
27. Zheng, X., Sauzeon, H.: Overview of the research on the application of virtual reality technology to neuropsychological assessment. Adv. Psychol. Sci. **18**(3), 511–521 (2010)
28. Roberts, A.C., Yeap, Y.W., Seah, H.S., Chan, E., Soh, C.K., Christopoulos, G.I.: Assessing the suitability of virtual reality for psychological testing. Psychol. Assess. **31**(3), 318–328 (2019)
29. Liu, T., Tang, Z.: Application of virtual reality technology in clinical psychology. In: 2020 International Conference on Computer Information and Big Data Applications (CIBDA) (2020)
30. Vasser, M., Kängsepp, M., Kilvits, K., Kivisik, T., Aru, J.: Virtual reality toolbox for experimental psychology—research demo. In: IEEE Virtual Reality (VR), pp. 361–362 (2015)
31. Smith, S.A.: Virtual reality in episodic memory research: a review. Psychon. Bull. Rev. **26**(4), 1213–1237 (2019). https://doi.org/10.3758/s13423-019-01605-w
32. American Psychiatric Association: Diagnostic and statistical manual of mental disorders, 5th edn. (2013)
33. Cheng, Y., Huang, C.L., Yang, C.S.: Using a 3D immersive virtual environment system to enhance social understanding and social skills for children with autism spectrum disorders. Focus Autism Other Dev. Disabil. **30**(4), 222–236 (2015)

34. Horace, I., et al.: Enhance emotional and social adaptation skills for children with autism spectrum disorder: a virtual reality enabled approach. Comput. Educ. **117**, 1–15 (2018)
35. Bailey, B., Bryant, L., Hemsley, B.: Virtual reality and augmented reality for children, adolescents, and adults with communication disability and neurodevelopmental disorders: a systematic review. Rev. J. Autism Dev. Disord. 1–24 (2021). https://doi.org/10.1007/s40489-020-00230-x
36. Code, J., Clark-Midura, J., Zap, N., Dede, C.: The utility of using immersive virtual environments for the assessment of science inquiry learning. J. Interact. Learn. Res. **24**(4), 371–396 (2013)
37. Parong, J., Mayer, R.E.: Learning science in immersive virtual reality. J. Educ. Psychol. **110**(6), 785–797 (2018)
38. Makransky, G., Andreasen, N.K., Baceviciute, S., Mayer, R.E.: Immersive virtual reality increases liking but not learning with a science simulation and generative learning strategies promote learning in immersive virtual reality. J. Educ. Psychol. **113**(4), 719–735 (2021)
39. You, S.H., et al.: Virtual reality-induced cortical reorganization and associated locomotor recovery in chronic stroke: an experimenter-blind randomized study. Stroke **36**(6), 1166–1171 (2005)
40. Golomb, M.R., et al.: In-home virtual reality videogame telerehabilitation in adolescents with hemiplegic cerebral palsy. Arch. Phys. Med. Rehabil. **91**(1), 1–8 (2010)
41. Li, W., Chau, T., Lam-Damji, S., Fehlings, D.: The development of a homebased virtual reality therapy system to promote upper extremity movement for children with hemiplegic cerebral palsy. Technol. Disabil. **8**(3), 107–113 (2009)
42. Neumann, D.L., et al.: A systematic review of the application of interactive virtual reality to sport. Virtual Reality **22**(3), 183–198 (2017). https://doi.org/10.1007/s10055-017-0320-5
43. Farley, O.R.L., Spencer, K., Baudinet, L.: Virtual reality in sports coaching, skill acquisition and application to surfing: a review. J. Hum. Sport Exerc. **15**(3), 535–548 (2020)
44. Carlo, G., Randall, B.A.: The development of a measure of prosocial behaviors for late adolescents. J. Youth Adolesc. **31**, 31–44 (2002). https://doi.org/10.1023/A:1014033032440
45. Rizzo, A.A., Bowerly, T., Buckwalter, J.G., Klimchuk, D., Mitura, R., Parsons, T.D.: A virtual reality scenario for all seasons: the virtual classroom. CNS Spectr. **11**(1), 35–44 (2006)
46. Jeong, D.C., Feng, D., Krämer, N.C., Miller, L.C., Marsella, S.: Negative feedback in your face: examining the effects of proxemics and gender on learning. In: Beskow, J., Peters, C., Castellano, G., O'Sullivan, C., Leite, I., Kopp, S. (eds.) IVA 2017. LNCS, vol. 10498, pp. 170–183. Springer, Cham (2017). https://doi.org/10.1007/978-3-319-67401-8_19
47. Haley, E., Yaremych., Susan, P.: Tracing physical behavior in virtual reality: a narrative review of applications to social psychology. J. Exp. Soc. Psychol. **85**, 103845 (2019)
48. Erik, V., Price, B. J., Bradley, C.: Direct Effects of Virtual Environments on Users. Handbook of Virtual Environments: Design, Implementation, and Application, pp. 521–529 (2015)
49. Gandhi, R.D., Patel, D.S.: Virtual reality – opportunities and challenges. Int. Res. J. Eng. Technol. (IRJET) **5**(1), 482–490 (2018)
50. Riva, G.: Virtual Reality in Clinical Psychology. Reference Module in Neuroscience and Biobehavioral Psychology (2020)
51. Araiza-Alba, P., Keane, T., Beaudry, J.L., Kaufman, J.: Immersive virtual reality implementations in developmental psychology. Int. J. Virtual Real. **20**(2), 1–35 (2020)
52. Parish-Morris, J., et al.: Immersive virtual reality to improve police interaction skills in adolescents and adults with autism spectrum disorder: preliminary results of a phase I feasibility and safety trial. Annu. Rev. Cyberther. Telemed. **16**, 50–56 (2018)
53. Sahin, N.T., Keshav, N.U., Salisbury, J.P., Vahabzadeh, A.: Safety and lack of negative effects of wearable augmented-reality social communication aid for children and adults with autism. J. Clin. Med. **7**(8), 188 (2018)

54. Afsoon, A., Taha, S.S., Zahra, S., Peyman, R.H.: Effectiveness of virtual reality- based exercise therapy in rehabilitation: a scoping review. Inform. Med. Unlocked **24**, 100562 (2021)
55. Freeman, D., et al.: Virtual reality in the assessment, understanding, and treatment of mental health disorders. Psychol. Med. **47**(14), 2393–2400 (2017)
56. Hamilton, D., McKechnie, J., Edgerton, E., Wilson, C.: Immersive virtual reality as a pedagogical tool in education: a systematic literature review of quantitative learning outcomes and experimental design. J. Comput. Educ. **8**(1), 1–32 (2020). https://doi.org/10.1007/s40 692-020-00169-2

XR Hackathon Going Online: Lessons Learned from a Case Study with Goethe-Institute

Wiesław Kopeć[1,2], Kinga Skorupska[1,2(✉)], Anna Jaskulska[1,3],
Michał Łukasik[1], Barbara Karpowicz[1], Julia Paluch[1],
Kinga Kwiatkowska[1], Daniel Jabłoński[1], and Rafał Masłyk[1]

[1] Polish-Japanese Academy of Information Technology, Warsaw, Poland
kinga.skorupska@pja.edu.pl
[2] SWPS University of Social Sciences and Humanities, Warsaw, Poland
[3] Kobo Association, Warsaw, Poland

Abstract. In this article we report a case study of a Language and Culture-oriented transdisciplinary XR hackathon organized with Goethe-Institut. The hackathon was hosted as an online event in November 2020 by our University Lab in collaboration with Goethe-Institut as a follow-up to our previous co-organized event within our research group Living Lab. We have improved the formula of the event based on lessons learned from its previous edition. First, in one of the two hackathon tracks we provided the participants with a custom VR framework, to serve as a starting point for their designs to skip the repetitive early development stage. In cooperation with our partner, Goethe-Institut, we have also outlined best modern research-backed language-learning practices and methods and gathered them into actionable evaluation criteria.

Keywords: Hackathon · Virtual reality · XR · Education

1 Introduction and Related Works

Hackathons have great potential for sparking inspiring ideas and collaborations and they provide great informal learning opportunities for students [3,6,9] and other participants, for example older adults, who may join hackathons as experts [4] - however, this massive potential is rarely realized in full. According to a survey of 150 hackathon participants [1], they do not participate because they want to build a product (26%), rather they do it for learning (86%) and networking (82%), especially if it is a corporate event, where the primary motivation, both of the organizers and attendees is finding employment [11]. Therefore, hackathons ought to be constructed in a more user-centric, or participant-centric manner, with these predominant motivations in mind, especially in learning environments. For this reason, in this online event we tried an organizational approach different from our previous practice (Fig. 1).

C. Biele et al. (Eds.): MIDI 2021, LNNS 440, pp. 218–228, 2022.
https://doi.org/10.1007/978-3-031-11432-8_22

Fig. 1. Our social media cover image for the VR Hackathon

The online event we organized was a follow-up to an event co-organized with Goethe-Institut a year prior, which we described in a case study at CHI 2021 [5] together with a rich set of insights based on the way it was organized, conducted as well as the final projects and participants' impressions. As project lifespan, and thus learning, after hackathons is limited because of motivation, follow-up, project understanding, team-composition, technologies used as well as a myriad of other factors [7] we decided to address some of these issues in this second edition.

First of all, we have built a clear set of guidelines and instructions in collaboration with our domain-specific expert partner institution - the Goethe-Institut. The same guidelines served to evaluate the projects after their completion. Next, keeping the learning and inclusivity in mind, we have enabled the participants to request features and changes to the hackathon formula prior to its launch. This resulted in lowering a barrier to entry, as the potential participants requested an additional track to enable them to participate without the requirement of having a functional VR headset at home.

Therefore, we added the second track devoted to AR projects, without any explicit hardware or framework but with a suggested broad pro-environmental theme. In addition to this track, we had the default VR track we have previously planned, which used our pre-fabricated custom VR framework to enable the participants to easily skip the basic and repetitive programming requirements and make the choice of technologies and the related trade-offs a non-issue [10]. Finally, we provided opportunities to active and willing participants to develop the projects further under the mentorship of experts in the field of XR development. Such extended, post-event mentorship, going beyond even the active participation of the mentor in the teams' decision making processes [8], can be a valuable asset to the participants. The opportunity to pursue follow-up work is well-aligned with participants' prevailing motivation connected to learning new things [1].

Therefore, in this case study, which falls well within the recent trends of research-oriented hackathons [2], we report the analysis, further insights and lessons learned from organizing a language and culture-oriented XR hackathon with Goethe-Institut. We refer to the benefits of using a provided technological framework as well as an extensive set of expert guidelines and evaluation criteria, both of which solved some of the problems we have faced in the previous edition of the hackathon.

2   Online XR Hackathon Case Study Method and Tools

Fig. 2. Part of the organizing team during the hackathon's live streamed kick-off meeting

2.1   Key Information

The hackathon was hosted as an online event on the 28–29.11.2020 by the XR Lab of the Polish-Japanese Academy of Information Technology in collaboration with Goethe-Institut and Kobo Association as a follow-up to our previous co-organized event [5]. To launch and end the event as well as communicate key information we used MS Teams, which was familiar to our students their online classes were conducted on the same platform. During the hackathon participants communicated with each other and with mentors using a dedicated Discord server. Additionally, the launch and end of the hackathon, including final project presentations, were broadcast using Instagram stories (Fig. 2).

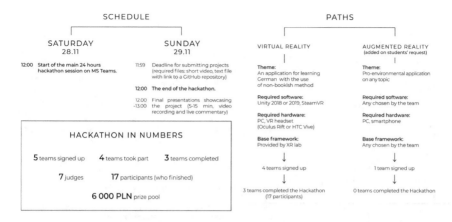

Fig. 3. Schematic overview of the online hackathon

2.2 Organization

AR Track vs VR Track with a Starting Framework. The hackathon allowed for a choice of two technology tracks - one for AR, and one for VR (Fig. 3). There were clear differences between them, as the AR track had a lower barrier to entry (no need to have a VR headset) and allowed the participants to use any technology of their choice. The project topic in this track had to fall within a broader pro-environmental theme. In the VR path the participants' task was to use a custom VR Framework, programmed by our XR Lab, to create an application for learning German with the use of modern immersive methods. The provided framework was dedicated for the Unity environment and consisted of the following parts:

- Scenes: MainMenu (interactive about and credits board, a player tutorial for teleportation, interacting with objects (doors, grabbing) and instructions for creating new scenes).
- Prefabs: Avatar (character animations for idle and talking) and Player (containing everything needed to navigate in VR, including hand scripting).
- Scripts: ControllerHints, HandManager (controllers), PickupManager (interacting with grabbable objects), Interactable (class for objects one can interact with using the laser), HighlightChanger (to highlight objects which are pointed at), HighlightMode, HighlightColors, Grabbable (for objects that can be picked up with the hand or the laser) and TeleportArea (to map where the player can teleport to).

The framework was accompanied by a comprehensive, 12-page manual outlining key functionalities, requirements and instructions for installing the development environment components, such as SteamVR, Unity 2018 or 2019 and the framework repository files.

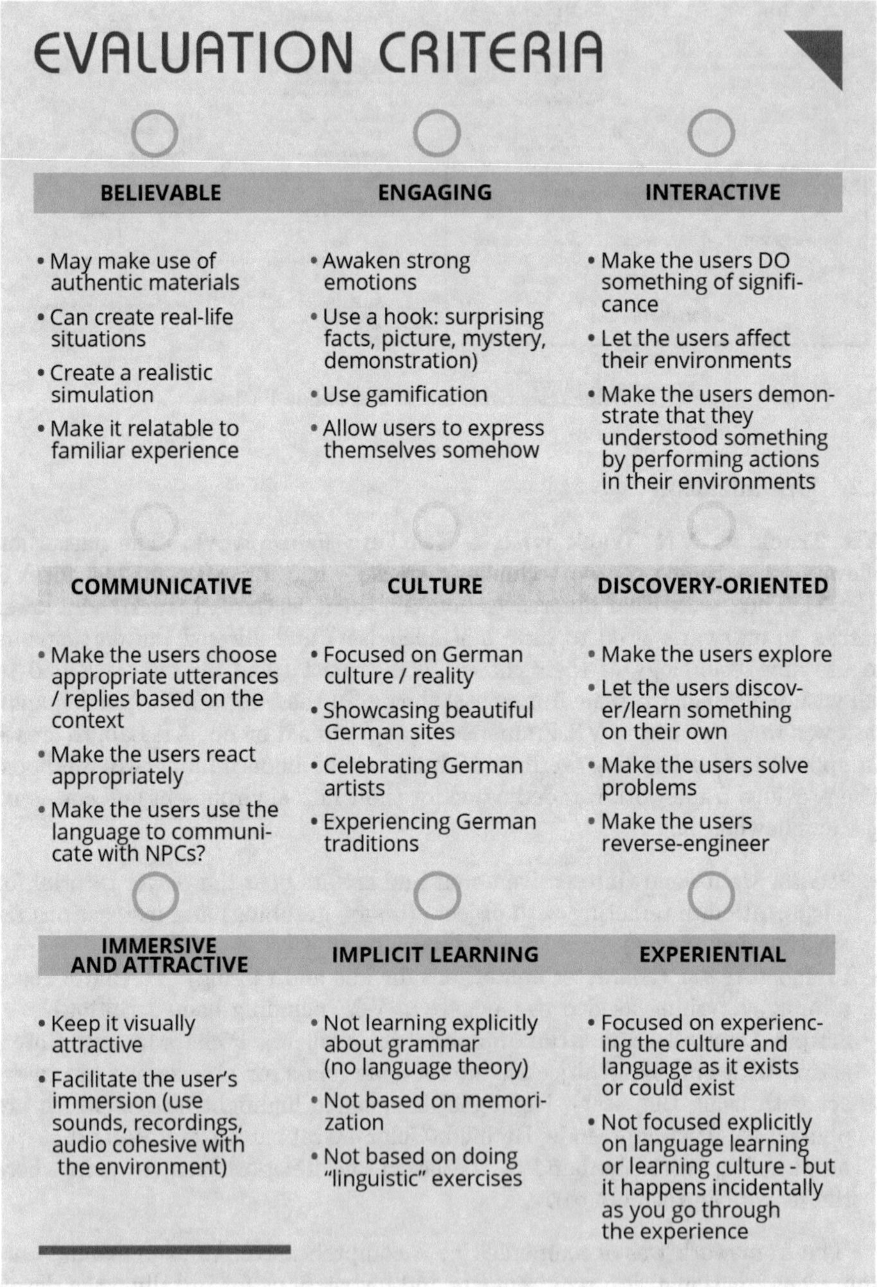

EVALUATION CRITERIA

BELIEVABLE

- May make use of authentic materials
- Can create real-life situations
- Create a realistic simulation
- Make it relatable to familiar experience

ENGAGING

- Awaken strong emotions
- Use a hook: surprising facts, picture, mystery, demonstration)
- Use gamification
- Allow users to express themselves somehow

INTERACTIVE

- Make the users DO something of significance
- Let the users affect their environments
- Make the users demonstrate that they understood something by performing actions in their environments

COMMUNICATIVE

- Make the users choose appropriate utterances / replies based on the context
- Make the users react appropriately
- Make the users use the language to communicate with NPCs?

CULTURE

- Focused on German culture / reality
- Showcasing beautiful German sites
- Celebrating German artists
- Experiencing German traditions

DISCOVERY-ORIENTED

- Make the users explore
- Let the users discover/learn something on their own
- Make the users solve problems
- Make the users reverse-engineer

IMMERSIVE AND ATTRACTIVE

- Keep it visually attractive
- Facilitate the user's immersion (use sounds, recordings, audio cohesive with the environment)

IMPLICIT LEARNING

- Not learning explicitly about grammar (no language theory)
- Not based on memorization
- Not based on doing "linguistic" exercises

EXPERIENTIAL

- Focused on experiencing the culture and language as it exists or could exist
- Not focused explicitly on language learning or learning culture - but it happens incidentally as you go through the experience

Fig. 4. Criteria used to evaluate the projects which were co-created by the language, art and IT experts from the jury and validated by Goethe-Institut prior to the hackathon.

	Believable	Engaging	Interactive	Comm.	Culture	Discovery	Immersive	Implicit	Experiential	Total
Team 1	5.5	5	5.5	5	6	4.5	6	5.5	6	49
SD	0.20	0.26	0.20	0.26	0.00	0.27	0.00	0.20	0.00	
Team 3	6	4	6	5	1.5	3	3.5	5.5	3	37.5
SD	0.00	0.26	0.00	0.26	0.27	0.32	0.20	0.20	0.32	
Team 4	5	2	5.5	2.5	2.5	1.5	2.5	5	2.5	29
SD	0.26	0.41	0.20	0.20	0.20	0.27	0.20	0.26	0.20	
ALL SD	0.19	0.37	0.16	0.30	0.38	0.34	0.30	0.21	0.33	

Fig. 5. Results of the jury evaluation of the team projects broken into specific criteria.

Project Presentation. The projects were submitted to each team's own folder on Google Drive. Each submission was to include a short video recording presenting the effect of work (gameplay, assets, features etc.) as well as a text file with a link to the project's repository on GitHub. The on-air 5 to 15-min presentation of the project, the video as well as live commentary by the participating teams.

Judging Criteria and Jury. For this hackathon, as the focus of our default track was culture- and language-oriented we have crated an extensive set of evaluation criteria with a description of each one, as shown in Fig. 4. Seven judges, experts in different areas including languages, research, IT, VR-development and art, could grant 0, 0.5 or 1 points to each project in each criterion depicted in Fig. 4. Additionally, each judge had the ability to grant up to 2 additional points to the projects they liked the most, provided they created their own criteria to justify their choice. The judging process was constructed this way to test the evaluation criteria chosen for this activity, as well as to enable the judges to form their own criteria, which could replace some of the existing criteria with low predictive value, or be added to the evaluation process in the future events.

3   Results

3.1   Teams and Tracks

The event was attended by 4 teams - including university and high school students. Each team consisted of at least 1 programmer and 1 artist to provide team diversity and transdisciplinarity, as per best practices [7]. The minimum number of people in a team was 3 and the maximum 6. Five teams signed up for the Hackathon, four took part in the competition, three of which completed the project with participant count of 6, 6 and 5 in each of them. Team number 2 resigned during the competition, as they could not finish the project to a satisfactory degree. All of the teams that successfully completed the hackathon were from the VR track. Here, one limitation of the online event was connected to granting appropriate space and time for ice-breaking activities and team formation facilitated by social preferences of the participants [12].

3.2   Judging Criteria

The results of the evaluation can be seen in Fig. 5. The categories of Believable, Interactive and Implicit, as explained in Fig. 4, failed to produce sufficiently different scoring results, with the standard deviation between all projects and all judges for these criteria at 0.19, 0.16 as well as 0.21. Similar results were obtained from interviews with the judges after the judging process, as they remarked that these criteria were either redundant (interactive, as VR by default is interactive), too general (believable) or easily achievable, thanks to prior instructions (Implicit Learning). Other criteria that the judges have specified themselves were: graphics (4 times), innovative idea (3 times) and one time for each: mood, humour, functionalities, effects and potential.

3.3   Winning Teams and Outcomes

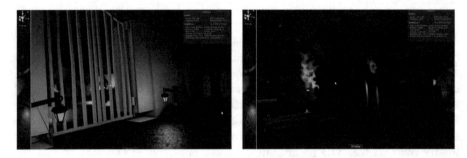

Fig. 6. Screenshots from the prototype made by team 1.

First Place. The winning project, created by team 1, was inspired by the atmosphere of traditional German outdoor festivals. This project was the most developed visually, relying on a visually-pleasing environment with atmospheric lightning, ambient music and custom models. The scene takes place in a night setting, in a park area where a festival is located. The player starts out in front of a gate leading into the festival grounds, surrounded by trees and warm lightning from torches and fireplaces. After interacting with the gate it opens, showing an alley with multiple tents and characters in fairytale-like costumes consisting of a cloak and a mask. Inside each tent there is a different minigame based on traditional childhood games widespread in Germany, such as Topfschlagen (heat-cold), Ein, Zwei, Drei... Halt! (Baba Yaga is watching) and Feuer-Wasser-Sturm-Blitz (fire-water-air-ground). The avatars instruct the player in German on how to proceed with each game; there are no subtitles to focus on improving listening competence. After finishing all games, all the NPCs move to a big bonfire located at the end of the alley, where an effigy is burned. The team received 49 points from the pre-determined evaluation criteria, as well as additional 3.5 points for graphics, 2.5 points for the idea, 1 for the mood, 1 for effects, and 0.5 for the potential (Fig. 6).

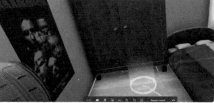

Fig. 7. Screenshots from the prototype made by team 3.

Second Place. The premise used for this project (by the team number 3) was a situation when an angry German mother is forcing the user to perform household chores, shouting the instructions in German. The environment conveyed the local atmosphere in a humorous way, with the mother dressed in traditional German apparel and over-the-top decorations in the interior (such as bedsheets with the German national flag or Rammstein poster on the wall). When picking up each object, its name was displayed in German, allowing the user to check the vocabulary. The quality of performance influenced the behaviour of the mother avatar, who reacted verbally in response to the user's interactions with objects, saying sentences such as "what are you doing?!" in the case of mistakes or "that's my son!" when the task was done correctly. The environment resembled an apartment with multiple chores to perform in different rooms. In the bedroom, the user had to pick up objects from the floor and place it in appropriate locations, e.g. dirty clothing in a clothing basket. In the kitchen there were dirty dishes to wash using a sponge. The environment was fully functional and interactive, with features such as a script allowing dirty clothing to spawn in a randomized way. The team grounded the application in a mnemonic device of associating information with an unusual or weird situation, which facilitates memorizing. They also presented ideas for the future development of the app, listing possible additional functionalities and further improvements. The team received 37.5 points from the pre-determined evaluation criteria, as well as 1 for humour, 1 for functionalities, and 0.5 for potential (Fig. 7).

Fig. 8. Screenshots from the prototype made by team 4.

Third Place. Team number 4 decided to make a game about cooking national German dishes (cooked sausage with potatoes). They created a scenario described as "You live in a cozy house near Munich, you're expecting friends for dinner tonight. Cook the best sausages with potatoes according to your great-grandmother's recipe". They created a virtual kitchen in which the user had to accomplish the task of cooking traditional German sausages with potatoes according to instructions in German displayed above the kitchen worktop. Next to instructions there was a timer counting down the time left to finish each stage. Ingredients had to be found inside a fridge, put on a frying pan and into a pot (which needed to be filled with water from the tap) placed on the stove and served on a platter. Although it was not clearly visible in the prototype due to the time constraints, the environment was intended to imitate a traditionally furnished German kitchen and convey the atmosphere of local culture. The team intended to extend the simulation with other recipes from the German cuisine. The team received 29 points from the pre-determined evaluation criteria and no additional points (Fig. 8).

4   Discussion and Conclusions

In this article we reported a case study of a Language and Culture-oriented VR hackathon with Goethe-Institut which was conducted online. Overall, in comparison to our previous Hackathon, the projects presented during this edition were more advanced, allowing for greater interaction and immersion. We improved the formula of the event based on lessons learned from its previous edition and based on this experience we offer additional considerations:

1. First, we provided the participants with the **ability to add their own technology track** to mitigate one of the biggest barriers related to having specialized VR equipment at hand. Yet, in our case, granting the participants more freedom did not work. The team in this track, despite having suggested this track themselves, did not finish their project, as likely they lacked the starting point provided by a pre-defined framework and limiting technology options.

2. Next, for our default track we have created **a custom VR framework, to serve as a starting point for the participants' designs to skip the repetitive early development stage.** Based on our pre-hackathon experience and questions we recommend to add a test-run of the said framework at least one day before the hackathon with a small "quest" to complete involving all the technical skills needed to take part in the hackathon. The results of the quest project would not count towards the hackathon scoring. We are also considering developing the framework further based on participant feedback, especially to include more ready-made scripts to enable a greater range of interactions, without limiting the participants' creativity by providing sets of assets directed at a certain interpretation of the tasks. In general, the use of programming frameworks is widespread in professional projects - therefore

we recommend this as a good practice during hackathons. Hackathon organizers, thanks to such provided framework, could better match the direction of participants' efforts to the hackathon goal.

3. Finally, for this hackathon we outlined best modern research-backed language-learning practices and methods and gathered them into **actionable evaluation criteria to ensure better understanding of the goal**. Whenever expert knowledge and practices are expected to be exhibited by non-professionals, such as in this case philology and methodology studies, it is advisable to briefly and clearly state the scope and expectations derived from subject expertise. In our case, we evaluated our criteria against their performance and have come up with an improved set for this specific purpose. The best criteria were: engagement, culture, discovery, experiential, immersive and communicative as explained in Fig. 4 – to these criteria we would add: graphics, innovative idea as well as potential for the project development.

We recommend the steps of providing a starting framework and extending expert evaluation criteria, to encourage the creation of more advanced projects, as well as to improve the experience of hackathon participation for the teams - which in general thanks to these practices which allow them to focus their creativity within a smaller range of possibilities, produce much better results. **These are especially important for online events, where direct communication within teams as well as with experts and mentors may be more difficult - as such, then these solutions provide an easy and common reference point.**

Acknowledgments. We would like to thank the many people and institutions gathered together by the distributed Living Lab Kobo and HASE Research Group (Human Aspects in Science and Engineering) for their support of this research. In particular, the authors would like to thank the members of XR Lab Polish-Japanese Academy of Information Technology and Emotion-Cognition Lab SWPS University as well as other HASE member institutions.

References

1. Briscoe, G., Mulligan, C.: Digital innovation: the hackathon phenomenon. Creativeworks London Working Paper **1**(6), 13 (2014)
2. Falk Olesen, J., Halskov, K.: 10 years of research with and on hackathons. In: Proceedings of the 2020 ACM Designing Interactive Systems Conference, DIS 2020, pp. 1073–1088. Association for Computing Machinery, New York (2020). https://doi.org/10.1145/3357236.3395543
3. Fowler, A.: Informal stem learning in game jams, hackathons and game creation events. In: Proceedings of the International Conference on Game Jams, Hackathons, and Game Creation Events, GJHGC 2016, pp. 38–41. Association for Computing Machinery, New York (2016). https://doi.org/10.1145/2897167.2897179
4. Kopeć, W., Balcerzak, B., Nielek, R., Kowalik, G., Wierzbicki, A., Casati, F.: Older adults and hackathons: a qualitative study. In: Proceedings of the 40th International Conference on Software Engineering, ICSE 2018, pp. 702–703. Association for Computing Machinery, New York (2018). https://doi.org/10.1145/3180155.3182547

5. Kopeć, W., et al.: VR hackathon with Goethe Institute: lessons learned from organizing a transdisciplinary VR hackathon. Association for Computing Machinery, New York (2021). https://doi.org/10.1145/3411763.3443432
6. Nandi, A., Mandernach, M.: Hackathons as an informal learning platform. In: Proceedings of the 47th ACM Technical Symposium on Computing Science Education, SIGCSE 2016, pp. 346–351. Association for Computing Machinery, New York (2016). https://doi.org/10.1145/2839509.2844590
7. Nolte, A., Chounta, I.A., Herbsleb, J.D.: What happens to all these hackathon projects? Identifying factors to promote hackathon project continuation. Proc. ACM Hum. Comput. Interact. 4(CSCW2), 1–26 (2020). https://doi.org/10.1145/3415216
8. Nolte, A., Hayden, L.B., Herbsleb, J.D.: How to support newcomers in scientific hackathons - an action research study on expert mentoring. Proc. ACM Hum. Comput. Interact. 4(CSCW1), 1–23 (2020). https://doi.org/10.1145/3392830
9. Prieto, M., Unnikrishnan, K., Keenan, C., Saetern, K.D., Wei, W.: Designing for collaborative play in new realities: a values-aligned approach. In: 2019 IEEE Games, Entertainment, Media Conference (GEM), New Haven, CT, USA, pp. 1–4. IEEE (2019). https://doi.org/10.1109/GEM.2019.8811545
10. Richterich, A.: Hacking events: project development practices and technology use at hackathons. Convergence 25(5–6), 1000–1026 (2019)
11. Ru, A., Khosmood, F.: Hackathons for workforce development: a case study. In: International Conference on Game Jams, Hackathons and Game Creation Events 2020, ICGJ 2020, pp. 30–33. Association for Computing Machinery, New York (2020). https://doi.org/10.1145/3409456.3409462
12. Trainer, E.H., Kalyanasundaram, A., Chaihirunkarn, C., Herbsleb, J.D.: How to hackathon: socio-technical tradeoffs in brief, intensive collocation. In: Proceedings of the 19th ACM Conference on Computer-Supported Cooperative Work and Social Computing, CSCW 2016, pp. 1118–1130. Association for Computing Machinery, New York (2016). https://doi.org/10.1145/2818048.2819946

Intergenerational Interaction with Avatars in VR: An Exploratory Study Towards an XR Research Framework

Barbara Karpowicz[1,5] (ID), Rafał Masłyk[1,5] (ID), Kinga Skorupska[1,3,5(✉)] (ID),
Daniel Jabłoński[1,5] (ID), Krzysztof Kalinowski[1,5] (ID), Paweł Kobyliński[4,5] (ID),
Grzegorz Pochwatko[2,5] (ID), Monika Kornacka[3,5] (ID), and Wiesław Kopeć[1,3,5] (ID)

[1] XR Lab, Polish-Japanese Academy of Information Technology, Warsaw, Poland
{karpowicz.b,kinga.skorupska,kopec}@pja.edu.pl,
kinga.skorupska@pjwstk.edu.pl
[2] VR and Psychophysiology Lab, Institute of Psychology Polish Academy of
Sciences, Warsaw, Poland
[3] Emotion Cognition Lab, SWPS University of Social Sciences and Humanities,
Warsaw, Poland
[4] Laboratory of Interactive Technologies, National Information Processing Institute,
Warsaw, Poland
[5] Kobo Association Living Lab and HASE Research Group, Warsaw, Poland

Abstract. The dynamic development of solutions in the field of virtual
and augmented reality poses challenges to designers. These challenges
relate to both technical conditions, including hardware capabilities and
software solutions, as well as psychophysical constructs conditioning the
end users' reception of the generated multimedia message. One of the key
elements of the virtual and augmented reality experience is the interaction with the system through a virtual agent represented by an avatar, i.e.
a reflection of the image of a participant in the virtual world, carrying on
a conversation with the user. This paper presents a proposed software
and hardware solution for conducting multifaceted research and comparative analysis of diverse interfaces and human-computer interaction
in virtual and augmented reality. In the course of this research, statistically significant results were obtained indicating differences in perception
between three types of virtual agents. Each of them represented by different avatars in a specially created research environment that allowed
to conduct usability tests under reproducible conditions to study user
interaction in virtual reality.

Keywords: Virtual and augmented reality · Human-computer
interaction · Virtual agent and avatar · User experience

1 Introduction and Related Works

Recent years have brought a dynamic growth of diverse approaches and solutions to the novel modes user interaction and interfaces in the field of Human-
Computer Interaction (HCI). One of the technologies undergoing significant

C. Biele et al. (Eds.): MIDI 2021, LNNS 440, pp. 229–238, 2022.
https://doi.org/10.1007/978-3-031-11432-8_23

evolution as new technical solutions and interfaces become available is Virtual Reality (VR), which challenges established paradigms of user interfaces, in particular well established WIMP paradigm (windows, icons, mouse, pointer).

This compels developers and academics to explore novel interfaces to facilitate effective human interaction with a three-dimensional virtual world, such as VR. There are multiple indicators of immersion in VR [14] in the field of applied psychophysiology [17], which may be used for the purpose of evaluating presence [5]. This aspect is key in evaluating interaction with avatars [2], or virtual agents [16], which are necessary in VR to engage people in social situations.

This study was a starting point for exploring various modes of voice interaction. It based on previous works of our XR Lab team in this field as a part of HASE research group (Human Aspects in Science and Engineering) by the Living Lab Kobo research activities on virtual reality rapid prototyping and development. These activities include end user engagement [10] and rapid content and software development [8] as well as alternative interfaces which were presented on major conferences, including CHI and also on previous MIDI conference, i.e. voice interfaces [7,11] and brain-computer interaction and interfaces (BCI) [9].

Therefore, the primary objective of this work is to propose a hardware and software solution that enables repeated experimental research of user interaction with agents equipped with various types of avatars. Another objective is to determine the differences in perception of various forms of avatars representing virtual agents in virtual reality. The main research hypothesis is that regardless of the visual depiction of the avatar, i.e. the virtual "person" giving the information, there are no variations in the user's perception of the identical content. In other words this research endeavour is in line with the concept of ecological validity.

The concept of ecological validity refers to experimental findings that can be generalized to real-life [1]. In research measuring emotional and cognitive processes, two approaches are often used – experimentally testing those processes in the laboratory or using retrospective recall with self-reported measures. Both of those methods are impairing the ecological validity of the study. First, laboratory settings are often very far from everyday life and thus the psychological processes measured in the lab might not fully reflect the everyday life of a given individual/group of individuals. Second, some of these processes, e.g. avoidance, can not be reliably measured through self-reported measures prone to retrospection biases. Thus, one of the main challenges in current research in the field of psychology and related disciplines is to assess and test psychological processes in the larger context of ecological validity, taking into account not only a given process but also the context of its development and maintenance [6]. Providing tools to study such psychological processes in the conditions of ecological validity is therefore a crucial research problem.

The results of this study coined the foundation for our XR framework for the development of advanced immersive environments and research tools providing ecological validity conditions with multimodal experimental data acquisition, including self-reported data (e.g. surveys) as well as objective psychophysiological data, related to eye movements, cardiac functions or skin conductance, described in the method section below. Therefore the results of this study paved the way for follow up studies and further research within the HASE group member labs, including Emotion Cognition Lab SWPS University and Institute of Psychology Polish Academy of Sciences.

2 Methods

2.1 Study Aims

To validate the study hypothesis while also evaluating the system's usability, the following research variants of the virtual agent interaction are compared (see Fig. 1). They are embedded in the same omnidirectional visual environment:

1. Avatar 1. High-fidelity model (rendered on the basis of photogrammetry) with scripted 3D animation,
2. Avatar 2. Video recording of a real person,
3. Avatar 3. VA (Voice Assistant), which is audio emitted from a virtual assistant model.

Fig. 1. From left to right: voice assistant case, high-fidelity photogrammetry model, video recording.

2.2 Mesures

As previously indicated, the study employed traditional research methodologies [3], both quantitative, including questionnaire surveys (conducted prior to, during, and after the VR session), and qualitative, in form of semi-structured interviews (prior to and after the VR session).

    These methods were validated using objective psychophysiological markers, specifically:

1. Eye Movement (EM) as a major sign of attention, measured by eye tracking,
2. Synchronized signals from auxiliary source, namely:
    (a) Cardiac function (PPG - PhotoPlethysmoGraphy, photoplethysmography, assessment of heart parameters based on blood flow analysis),
    (b) Changes in skin conductivity (EDA/GSR - ElectroDermal Activity, Galvanic Skin Response).

The automated measurement of the aforementioned psychophysiological indicators within the proposed approach (research framework) was utilized to generate objective measures for evaluating the reliability of reception of the presented content. The objective of such verification was to eliminate inconsistencies in declarative data that are caused by natural human factors and are inherent in evaluating human-computer interaction, such as: the Hawthorne effect, [12,13] which refers to the impact of the researcher's presence and implicit expectations on the subject's response, the desire to present a subject more proficiently than other subjects, and the possibility of obtaining insincere answers from the participants.

To test the research hypothesis, the results of the study participants' declarative responses were compared to psychophysiological data on several dimensions relevant to assessing the immersion quality of the user's interaction with virtual reality [5]. These dimensions include sense of immersion and co-presence, as well as the attribution of anthropomorphic features to agents, taking into account potential occurrence of the uncanny valley effect, which has been studied extensively in, and outside, of VR [15]. The last factor is especially pertinent when evaluating the quality of potential high-fidelity content, particularly humanoid avatar models [4].

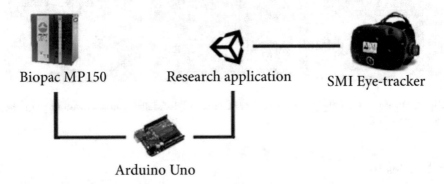

Biopac MP150   Research application   SMI Eye-tracker

Arduino Uno

Fig. 2. Schematic of the proposed research solution.

2.3   Research Application

The research conducted for this work resulted in the creation of the dedicated solution depicted in Fig. 2, which was subsequently validated through an empirical survey with users, as detailed later. The research solution consisted of:

1. Arduino - to mediate the Unity - Biopac communication.
2. Unity - with necessary prefabs such as: GazeObjectManager, The EyetrackerMasnager, SMI_CameraWithEyeTracking and SceneSwitcher

The following tools were utilized to develop the software required for the study: Unity, the Arduino IDE, MS Visual Studio, and the HTC iViewHMD software.

2.4 Research Flow

Process name	Description
Baseline	Gathering the data from the Biopac sensors without the headset to serve as a baseline for evaluating the psychophysiological data gathered during the experiment proper. The participant for 5 min is alone in a room, sitting in front of a black wall
Survey settings	The first scene after turning on the application visible only to the researcher. Here, enter the prefix of the result files for the test subject and the port number to which the Arduino is connected. Additionally, you can select the data simulation mode for the eye tracker to facilitate testing the application
Startup scene/calibration	The first scene visible to the subject. This is where the eye tracker is calibrated and the order of the scenes presented is implicitly selected
Preliminary survey (training, warm-up)	A scene that allows the subject to become familiar with the questionnaire interface. Additionally, it will serve to establish the subject's baseline mood
VR 360 scenes with an agent	The main scene of the application showing stages with different agents in VR: 3D animation, video recording and voice assistant in a random order for different participants
Follow-up questionnaire after each VR scene	Scene used for survey, after each stage with an assistant

2.5 Experimental Setup

The pilot experimental study was conducted in the Institute of Psychology of the Polish Academy of Sciences' VR Lab. IP PAN's VR Lab is equipped with the technological equipment fulfilling the requirements of the study, including a SMI eye tracker paired with a virtual reality headset, a system for psychophysiological assessments (Biopac), as well as statistical analysis capabilities and research hypothesis verification (Fig. 3).

2.6 Participants

The pilot study involved twenty-two Living Lab Kobo participants, including 18 from the experimental group, which included seniors over the age of 60, and 4 from the control group (under 50 years old). The experimental group consisted

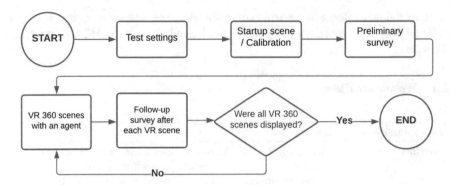

Fig. 3. Workflow of the survey application.

of 13 women and 9 men. The mean age for the entire study was 64.1 (standard deviation, SD = 15.52), with the experimental group averaging 70.8 (SD = 7.65) and the control group averaging 37.25 (SD = 8.31). The study's youngest participant was 23 years old, while the oldest was 90 years old. The median age was 66 years, with equals 68 in the experimental group and 41 in the control group. 22 sets of measurements were taken throughout the study, which included 18 sets of measurements from the experimental group and 4 sets of measurements from the control group (Fig. 4).

3   Results

The results of declarative (ex-ante, control, and ex-post questionnaires) and psychophysiological (EM, PPG, and EDA) tests conducted during the analyses revealed statistically significant differences in perception of avatars, supporting the rejection of the hypothesis that no differences in perception of different types of avatars representing virtual agents in virtual reality exist.

The results of participants' declarative responses to survey questions asked both before and after the study (on paper) and during the study: via a questionnaire module integrated into the research framework - were utilized to verify the research hypothesis. The questionnaire responses were examined in the context of psychophysiological data gathered using eye tracker EMs synchronized with Biopac signals (PPG and EDA/GSR).

Data analysis was conducted on several psychophysical dimensions identified in the formulation of the research problem that are relevant to assessing the immersion quality of a user's interaction with virtual reality, specifically: a sense of immersion in virtual reality, a sense of co-presence, attributing anthropomorphic characteristics to agents, Belief in Human Nature Uniqueness (BHNU), and the uncanny valley effect. BHNU had a particularly strong link with the experience of co-presence in scenes with a humanoid avatar, with a correlation

Fig. 4. Preparing the study participant at the VRLab IP PAS

coefficient of 0.57 for video footage and 0.44 for rendered avatar. This demonstrates that the produced avatar lacked the sense of its human features present in the image from the video clip.

Moreover, additional extensive analyses of the anthropomorphic qualities assigned to avatars revealed further evident and statistically significant differences in perceptions of avatars. Fewer participants attributed human characteristics to the VA avatar than to a video recording and rendered avatar. The perceived sense of co-presence was most prominent for the video, decreased in case of rendered avatar, and was lowest for the VA.

Additionally, statistically significant variations were discovered in evaluations of the uncanny valley dimension: the phenomenon was observed the most for rendered avatar and occurred the least for the video recording.

The findings shown above, which are based on questionnaire and psychophysiological data, are consistent with the information gathered from in-depth qualitative interviews, as well as the analysis of eye tracker data.

4   Discussion

The results of the pilot study conducted in XR Lab PJAIT in cooperation with Emotion Cognition Lab SWPS University and Virtual Reality and Psychophysiology Lab of the Institute of Psychology of the Polish Academy of Sciences were deemed very promising by members of the HASE research group. As a result, work on the provided solution will resume, and the framework will undergo further development. Further waves of the study are planned to confirm the hypotheses by expanding the number of the experimental group of seniors and the control group of younger individuals.

With these objectives in mind, it is worth noting that the configuration of the connection between Arduino and Biopac, which is critical for synchronizing psychophysiological signals and correlating them to declarative questionnaire responses, proved to be effective and sufficient. However, due to the nature of the basic Biopac module (electrically unbuffered diagnostic ports), a more secure solution utilizing a specialized Biopac module for digital communication (STM type) or the use of an additional installation galvanically separating the electrical signal, such as an optocoupler, is recommended for the future.

5   Conclusions

The solution presented in this paper was validated through an experimental research procedure with users, demonstrating its efficacy and utility in resolving the primary research problem, which is the evaluation of interactions in virtual reality via new interfaces in the form of virtual agents with a variety of avatars. The experiment demonstrates that the numerous psychological measures used to assess users' immersion in virtual reality reveal statistically significant variations in agents' and their avatars' perceptions. At the same time this study formed the basis for further work on the XR framework, which enables research teams to conduct XR experiments in the conditions of ecological validity, while at the same time verifying their qualitative findings through numerous psychophysiological measures. Such alignment of multimodal research measures in the immersive virtual reality enables the development of reproducible experiments providing more reliable, triangulated, results.

Acknowledgements. This study constitutes an example of a bottom-up participatory research initiative done in the spirit of transdisciplinary collaboration between scientists, practitioners and volunteers. It was conducted without a dedicated grant to further the understanding of key concepts in HCI in the context of immersive virtual environments (IVR) and it constitutes the birth of a dedicated framework for the development of immersive interactive VR and XR research tools.

Therefore, we would like to thank the many people and institutions gathered together by the Living Lab Kobo and HASE Research Group. First, we would like to thank all the members of HASE research group (Human Aspects in Science and Engineering) and Living Lab Kobo for their support of this research. In particular, the members of XR Lab Polish-Japanese Academy of Information Technology (PJAIT) and Emotion-Cognition Lab SWPS University (EC Lab) for controlling the experimental conditions and setup alongside with coordination and facilitation of the entire experiment, Kobo Association (special thanks to Anna Jaskulska) for supporting the construction of the framework incl. electronic engineering (Sebastian Zagrodzki), Living Lab Kobo community, especially older adults, for supporting recruitment and their participation in the lab studies, VR and Psychophysiology Lab of the Institute of Psychology Polish Academy of Sciences for the access to psychophysiology research tool, software and support for the lab experimental setup and conducting and the lab studies, as well as 3D Lab PJAIT (especially Jakub Tyszka, Roman Karowiec and Martyna Bihun from Krzysztof Kalinowski's team) for contributing 3D content and Laboratory of Interactive Technologies of National Information Processing Institute for supporting multi-modal data analysis.

References

1. Andrade, C.: Internal, external, and ecological validity in research design, conduct, and evaluation. Indian J. Psychol. Med. **40**(5), 498–499 (2018)
2. Baylor, A.L.: Promoting motivation with virtual agents and avatars: role of visual presence and appearance. Philos. Trans. Roy. Soc. B Biol. Sci. **364**(1535), 3559–3565 (2009)
3. Brannen, J.: Mixing methods: the entry of qualitative and quantitative approaches into the research process. Int. J. Soc. Res. Methodol. **8**(3), 173–184 (2005). https://doi.org/10.1080/13645570500154642
4. Cheetham, M., Suter, P., Jäncke, L.: The human likeness dimension of the "uncanny valley hypothesis": behavioral and functional MRI findings. Front. Hum. Neurosci. **5**, 126 (2011)
5. Dillon, C., Keogh, E.: Aroused and immersed: the psychophysiology of presence. Citeseer (2000)
6. Hayes, S.C., Hofmann, S.G., Wilson, D.S.: Clinical psychology is an applied evolutionary science. Clin. Psychol. Rev. **81**, 101892 (2020)
7. Jaskulska, A., Skorupska, K., Karpowicz, B., Biele, C., Kowalski, J., Kopeć, W.: Exploration of voice user interfaces for older adults—a pilot study to address progressive vision loss. In: Biele, C., Kacprzyk, J., Owsiński, J.W., Romanowski, A., Sikorski, M. (eds.) MIDI 2020. AISC, vol. 1376, pp. 159–168. Springer, Cham (2021). https://doi.org/10.1007/978-3-030-74728-2_15
8. Kopeć, W., et al.: VR hackathon with Goethe Institute: lessons learned from organizing a transdisciplinary VR hackathon. Association for Computing Machinery, New York (2021). https://doi.org/10.1145/3411763.3443432
9. Kopeć, W., et al.: Older adults and brain-computer interface: an exploratory study. Association for Computing Machinery, New York (2021). https://doi.org/10.1145/3411763.3451663
10. Kopeć, W., et al.: VR with older adults: participatory design of a virtual ATM training simulation. IFAC-PapersOnLine **52**(19), 277–281 (2019). https://doi.org/10.1016/j.ifacol.2019.12.110. https://www.sciencedirect.com/science/article/pii/S2405896319319457. 14th IFAC Symposium on Analysis, Design, and Evaluation of Human Machine Systems, HMS 2019
11. Kowalski, J., et al.: Older adults and voice interaction: a pilot study with Google home. In: Extended Abstracts of the 2019 CHI Conference on Human Factors in Computing Systems, CHI EA 2019, pp. 1–6. Association for Computing Machinery, New York (2019). https://doi.org/10.1145/3290607.3312973
12. Macefield, R.: Usability studies and the Hawthorne Effect. J. Usability Stud. **2**(3), 145–154 (2007)
13. McCarney, R., Warner, J., Iliffe, S., van Haselen, R., Griffin, M., Fisher, P.: The Hawthorne Effect: a randomised, controlled trial. BMC Med. Res. Methodol. **7**(1), 30 (2007). https://doi.org/10.1186/1471-2288-7-30

14. Pugnetti, L., Meehan, M., Mendozzi, L.: Psychophysiological correlates of virtual reality: a review. Presence Teleoper. Virtual Environ. **10**(4), 384–400 (2001)
15. Seyama, J., Nagayama, R.S.: The uncanny valley: effect of realism on the impression of artificial human faces. Presence Teleoper. Virtual Environ. **16**(4), 337–351 (2007). https://doi.org/10.1162/pres.16.4.337
16. Wang, I., Smith, J., Ruiz, J.: Exploring virtual agents for augmented reality. In: Proceedings of the 2019 CHI Conference on Human Factors in Computing Systems, pp. 1–12 (2019)
17. Wiederhold, B.K., Rizzo, A.: Virtual reality and applied psychophysiology. Appl. Psychophysiol. Biofeedback **30**(3), 183–185 (2005)

Multisensory Representation of Air Pollution in Virtual Reality: Lessons from Visual Representation

Grzegorz Pochwatko[1]([✉])[iD], Justyna Świdrak[1,2][iD], Wiesław Kopeć[3][iD], Zbigniew Jędrzejewski[1][iD], Agata Feledyn[1][iD], Matthias Vogt[4][iD], Nuria Castell[4][iD], and Katarzyna Zagórska[5][iD]

[1] Institute of Psychology, Polish Academy of Sciences, Warsaw, Poland
{gp,jswidrak}@psych.pan.pl
[2] August Pi & Sunyer Biomedical Research Institute, Barcelona, Spain
[3] Polish-Japanese Academy of Information Technology, Warsaw, Poland
[4] Norwegian Institute for Air Research, Kjeller, Norway
[5] Faculty of Economic Sciences, University of Warsaw, Warsaw, Poland

Abstract. The world is facing the problem of anthropogenic climate change and air pollution. Despite many years of development, already established methods of influencing behaviour remain ineffective. The effect of such interventions is very often a declaration of behaviour change that is not followed by actual action. Moreover, despite intensive information campaigns, many people still do not have adequate knowledge on the subject, are not aware of the problem or, worse, deny its existence. Previous attempts to introduce real change were based on providing information, persuasion or visualisation. We propose the use of multi-sensory virtual reality to investigate the problem more thoroughly and then design appropriate solutions. In this paper, we introduce a new immersive virtual environment that combines free exploration with a high level of experimental control, physiological and behavioural measures. It was created on the basis of transdisciplinary scientific cooperation, participatory design and research. We used the unique features of virtual environments to reverse and expand the idea of pollution pods by Pinsky. Instead of closing participants in small domes filled with chemical substances imitating pollution, we made it possible for them to freely explore an open environment - admiring the panorama of a small town from the observation deck located on a nearby hill. Virtual reality technology enables the manipulation of representations of air pollution, the sensory modalities with which they are transmitted (visual, auditory, tactile and smell stimuli) and their intensity. Participants' reactions from the initial tests of the application showed that it is a promising solution. We present the possibilities of applying the new solution in psychological research and its further design and development opportunities in collaboration with communities and other stakeholders in the spirit of citizen science.

Keywords: Virtual reality · Air pollution · Participatory design · User experience · Art and science

C. Biele et al. (Eds.): MIDI 2021, LNNS 440, pp. 239–247, 2022.
https://doi.org/10.1007/978-3-031-11432-8_24

1   Theoretical Context

1.1   Introduction

Reduction of air pollution is an urgent challenge that European societies and governments have to face. Despite policies in place, yearly air quality improvements fall short to Europe's zero pollution ambition. Worldwide, around 90% of people breathe polluted air (97% of Poles - six out of ten most polluted cities of Europe are located in Poland). Over the past 6 years, air pollution levels have remained high and stable, way above the levels recommended by the World Health Organization. Air pollution leads to increased mortality, one in five (20%) natural deaths are attributed to air pollution [14,27]. The destructive consequences go beyond death and respiratory symptoms: anxiety, dementia, missed work, increased harm of Covid etc. Air pollution is largely a behavioural problem - it can undeniably be connected with human actions (burning solid fuels), not industrial sources. Air pollution has local, close to source, consequences: it directly affects one's own health and that of the neighbours, as well as the direct environment (fauna, flora, urban surroundings). It creates a microsystem, which can be thoroughly monitored, measured and modelled in transdisciplinary cooperation between social and atmospheric scientists. However, before we start with the field studies, the carefully designed virtual environment will allow us to understand the behavioural processes already in place and build evidence-based models with potential for replication in real contexts and places. The role of modern virtual reality technologies is to make testing solutions cheaper and more effective.

In the following sections: we will describe what the problem with changing air pollution behavior is and how virtual reality can help; we will discuss what different types of risk communication are and how you can use multiple senses to make it more effective; we will present studies showing psychophysiological responses to risk messages and how this can be used in a research application to select the most effective air pollution messages (but not only as these responses are quite universal); we will characterize the basics and the process of creating a multisensory virtual environment, taking into account the participatory process; finally, we share our first experiences of testing an environment containing visual stimuli.

1.2   The Problem

Air pollution is an invisible killer. Our aim is to effectively communicate the health and environmental threat. We will use the potential of virtual reality technologies to find, test and implement novel solutions that influence the way people envision and battle the problem of air pollution. We investigate how multisensory virtual experience and pollution visualisation impact environmental attitudes and behaviours. Before being introduced in the field studies, each form of multisensory virtual experience has to be tested in virtual reality labs. In the lab-experiments we will stimulate multiple senses (vision, hearing, touch, and smell) in a controlled and replicable manner, and gather high quality reliable data about people's reactions to sensory stimuli.

1.3   Risk Communication

Visualisations and sound signals designed to build awareness of potential dangers are encountered daily and serve the purpose of saving people from harming themselves or others [18]. Examples of risk communication design include road signs, labels of toxic or poisonous substances, tobacco products' packaging, alert sounds. They need to be clearly and instantly distinguishable and understandable despite the receiver's age, culture or graphic literacy [5, 18, 24]. Fundamental elements of risk communication design ought to evoke mutual reactions. Inspiration often comes from signals observed in nature e.g. colours of poisonous frogs or venomous snakes [9, 17]. Research in the field focuses on principle elements of communication such as shapes, colours or sounds that evoke attention or disgust [13, 18, 24, 28]. Specifically, exposure to visualisations of shapes resembling the letter "V" evoke reaction in amygdala: part of the brain responsible for fight-or-flight defensive response [16]. Example of using other senses could be alerting drivers with haptic feedback [26]. A use of risk communication may be encountered almost everywhere, from supermarket shelves to the deep forest. Virtual Reality technology has a great potential of implementing risk communication signals, but research using it has a relatively short history, as only the last dozen or so years has brought the increasing popularity of this technology and its use outside the laboratory.

1.4   Neurophysiological Response to Risk Signals

A strong multimodal response increases the likelihood of the effectiveness of the risk signals. Testing them in VR gives the opportunity to check not only the participants' declarations, but also the use of more objective measurements - psychophysiological reactions. Millisecond measurement precision and synchronization with external medical-class equipment allow you to record any signals, but it is worth selecting the most promising ones.

Threat-related signals induce strong neurophysiological responses associated with both emotional and attentional processes. Affective neuroscience has identified the amygdala as the primary neural site responding to danger-related stimuli [20]. Through its bidirectional connections with sensory regions, the amygdala has been proposed to enhance sensory processing of emotionally relevant stimuli [23], as evident by increased activity in the visual cortex [8, 22] and auditory cortex [12]. In addition to subcortical areas involved in processing of affective stimuli, the prefrontal cortex (PFC) has been implicated in emotion regulation [2, 7]. Increased activity of the PFC was observed upon presentation of fearful stimuli across different modalities in experiments using functional magnetic resonance imaging (fMRI) [3], functional near-infrared spectroscopy (fNIRS) [11, 19] as well as electroencephalography (EEG) [8].

Risk signals, perhaps via emotional processes, have been found to effectively capture and hold attention [15, 23]. Selective attention to threat is exerted through the interaction between bottom-up (perceptual) and top-down (cognitive) processes [6]. At the level of perception, threat-related stimuli are detected more efficiently [10] and attract the gaze with a higher frequency [4] than neutral stimuli. Cognitive processes modulating attention are linked to activity in the frontal and parietal cortices [1]. In summary, neuroscientific findings on risk-related signals have pointed to the role of amygdala, prefrontal cortex and sensory cortices associated with emotional processing and to the frontal and parietal regions recruited during selective attention to threat.

2 Construction of the Environment

Considering the above, there is a need to construct a virtual environment for use in immersive virtual reality that will enable: 1) testing multi-sensory representations of air pollution (visual, auditory, tactile and smell); 2) registration of emotional, attention and behavioural responses of participants; 3) millisecond precision; 4) communication with external research equipment. The use of a participatory approach implies a test with the participation of end-users, i.e. residents of small towns struggling with the problem of air pollution caused by residential and water heating.

2.1 Inspiration - Pollution Pods

The design of the virtual environment was inspired by the artistic installation by Michael Pinsky - the pollution pods, which he himself describes as an answer to the challenge of "representing the invisible". He tried to apply the knowledge from the fields of environmental psychology, empirical aesthetics, and activist art. Pollution Pods is a sensorial experience created in five domes connected with each other to form a ring. Within each dome artist with the help of scientists recreated the air quality of five global cities. The visitors (volunteers from Trondheim and London, as the installation was displayed in public space) passed through increasingly polluted cells. The effect of pollution pods on visitors was mixed. E.g. reported intentions to act were strong and increased after participation, but were not followed by actual behaviour (participants did not track their climate change emissions afterwards). Nevertheless, it was concluded that environmental art can be useful for environmental communication [21, 25].

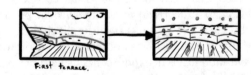

Fig. 1. First draft of the virtual environment - stimulus intensity

2.2   The Multisensory Virtual Environment

The need to reliably recreate the smell of polluted air forced the author of the "pollution pods" to use small, tightly sealed domes. This necessary procedure, however, made the whole experience somehow artificial and distant from actually being in Trondheim, London or New Delhi and breathing the local breeze. Participants were not able to experience being in the city space, characteristic sights and sounds, disturbance of the aerial perspective due to pollution. The use of VR overcomes these limitations and provides additional opportunities to enhance and go beyond reality. Instead of closing participants in small domes filled with chemical substances imitating pollution, we made it possible for them to freely explore an open environment - admiring the panorama of a small town from the observation deck located on a nearby hill. The deck was divided into five separate spaces - terraces. They were located around the hill in such a way that one could see a fragment of the next one from each of them, but the view from it was obstructed by the hill (Fig. 2). From each of the terraces one could see a fragment of the town, houses, trees and characteristic buildings. Thanks to this, the participants could experience a slightly different view from each of the terraces. These views, however, were perceptually similar and complex to the same degree, which meant that the observed changes in behavior on each of the terraces could only be the result of experimental manipulation, not different environmental conditions. The type and intensity of the air pollution stimuli could be freely adjusted on individual terraces and changed over time. An example scenario may be the increasing concentration of PM2.5 and PM10 particles magnified so that they become visible to the naked eye (Fig. 1).

An additional measure to increase the realism of the experience was to adjust the size of the terrace to the size of the laboratory. Thanks to this, with the use of a wireless VR set, participants could freely explore the terraces. High barriers at an appropriate distance were put in place to prevent collisions with the walls which could cause injuries as well as breaks in presence. Additionally, the hill behind the barriers descended gently to prevent unpleasant sensations caused by the fear of heights. Comfortable movement in the environment is also possible thanks to short, standardized training before starting the actual simulation.

The use of VR technology made it possible to track the behavior of participants and their reactions to various representations of air pollution. The dynamics of movement in space can be followed thanks to the constant registration of the position and rotation of the participant's body. In addition, eye movements, dilation of the pupils and the id of objects on which the gaze falls at a given moment are recorded. Thanks to markers sent to the external apparatus via parallel port, synchronization with measures of physiological reactions is possible. The modular form of the VR application ensures its scalability. The number and type of stimulation the participants are to experience may be manipulated in any way. The introduction of the possibility of answering the questions of the questionnaires without removing the HMD additionally improves the usability and does not interrupt the feeling of immersion in the virtual world.

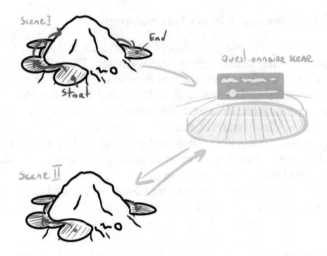

Fig. 2. First draft of the virtual environment - scenes

2.3   The Participatory Activities and Design Process

The virtual environment idea was one of the topic of participatory design and research workshops during the project kick-off with local community and stakeholders. It was a part of the citizen science approach with direct potential end users' involvement. At the next step the early-stage prototype was presented to volunteers, inhabitants of the town of Myszków, which is on the list of 50 Polish cities with the most polluted air. They are representatives of potential recipients of solutions developed on the basis of laboratory tests in VR. Citizens and various project stakeholders of Myszków participated in the Family Ecological Picnic organized jointly with the local authorities of the town and non-governmental organizations. The VR experience was one of the highlights of the event. Due to the limitations related to the COVID19 pandemic, participants experienced the VR environment individually with all necessary precautions and in an open-air setup on a stage in a local activity park. During and after the experience, they were able to share their impressions and provide feedback to team members. An additional group were volunteers who participated in the laboratory experiment. They had the opportunity to record their comments during the experiment. Below are some representative statements from VR experience participants:

«*Is it really so much of this [particulate matter] floating in the air now? But today the air seems clear, how come?*»

«*It's scary...it is all around*»

«*Objects floating in the air made me feel sad and somehow unsafe. They resembled insects, but they weren't them. I felt uncomfortable, especially as there were more and more of them.*»

«*The air was full of some black creatures. As it got more and more of them, they looked like locusts. It wasn't pleasant. But the next visuals were very pleasant, I like such atmosphere very much. I could stay in them for a long time.*»

«*I would like to always see the world this way.*»

Direct involvement of end users delivered valuable and immediate feedback that was very insightful for the next stages of the development of our immersive multisensory VR experience. Therefore, it was also an interesting example of citizen science - as it enabled the participants to shape the development of our immersive environment and research tool to be used for further VR lab studies.

3 Conclusions

The interdisciplinary approach to the project, both since its conception and inspiration from artistic installations, through research involving psychological, economic, social and technological dimensions is a firm step towards stronger collaboration between disciplines. Such collaboration, which may be dubbed as transdisciplinary, when paired with the involvement of local communities and other stakeholders, in the spirit of citizen science, is a very promising way to address the complex and wicked problems of today's world, such as environmental challenges, or, as in our case, air-pollution.

Acknowledgment. This project received funding from #eeagrants NCN IDEALAB grant no. 2019/35/J/HS6/03166. We would like to thank our participants, volunteers from VRLab IP PAN panel and citizens of Myszków, Poland, who took part in our studies.

References

1. Armony, J.L., Dolan, R.J.: Modulation of spatial attention by fear-conditioned stimuli: an event-related fMRI study. Neuropsychologia **40**(7), 817–826 (2002)
2. Banks, S.J., Eddy, K.T., Angstadt, M., Nathan, P.J., Phan, K.L.: Amygdala-frontal connectivity during emotion regulation. Soc. Cogn. Affect. Neurosci. **2**(4), 303–312 (2007)
3. Bermpohl, F., et al.: Dissociable networks for the expectancy and perception of emotional stimuli in the human brain. Neuroimage **30**(2), 588–600 (2006)
4. Bradley, B.P., Mogg, K., Millar, N.H.: Covert and overt orienting of attention to emotional faces in anxiety. Cogn. Emot. **14**(6), 789–808 (2000)
5. Bresciani, S., Eppler, M.J.: The pitfalls of visual representations: a review and classification of common errors made while designing and interpreting visualizations. SAGE Open **5**(4), 2158244015611451 (2015)
6. Corbetta, M., Shulman, G.L.: Control of goal-directed and stimulus-driven attention in the brain. Nat. Rev. Neurosci. **3**(3), 201–215 (2002)
7. Davidson, R.J., Putnam, K.M., Larson, C.L.: Dysfunction in the neural circuitry of emotion regulation-a possible prelude to violence. Science **289**(5479), 591–594 (2000)
8. DeLaRosa, B.L., et al.: Electrophysiological spatiotemporal dynamics during implicit visual threat processing. Brain Cogn. **91**, 54–61 (2014)
9. Donath, J.: Signals, cues and meaning. Unpublished Manuscript. Massachusetts Institute of Technology, Cambridge, MA (2007). http://smg.media.mit.edu/classes/IdentitySignals06/SignalingDraft.pdf

10. Fox, E., Lester, V., Russo, R., Bowles, R., Pichler, A., Dutton, K.: Facial expressions of emotion: are angry faces detected more efficiently? Cogn. Emot. **14**(1), 61–92 (2000)
11. Glotzbach, E., Mühlberger, A., Gschwendtner, K., Fallgatter, A.J., Pauli, P., Herrmann, M.J.: Prefrontal brain activation during emotional processing: a functional near infrared spectroscopy study (fNIRS). Open Neuroimaging J. **5**, 33 (2011)
12. Grandjean, D., et al.: The voices of wrath: brain responses to angry prosody in meaningless speech. Nat. Neurosci. **8**(2), 145–146 (2005)
13. Kemp, D., Niederdeppe, J., Byrne, S.: Adolescent attention to disgust visuals in cigarette graphic warning labels. J. Adolesc. Health **65**(6), 769–775 (2019)
14. Khomenko, S., et al.: Premature mortality due to air pollution in European cities: a health impact assessment. Lancet Planetary Health **5**(3), e121–e134 (2021)
15. Koster, E.H., Crombez, G., Van Damme, S., Verschuere, B., De Houwer, J.: Does imminent threat capture and hold attention? Emotion **4**(3), 312 (2004)
16. Larson, C.L., Aronoff, J., Sarinopoulos, I.C., Zhu, D.C.: Recognizing threat: a simple geometric shape activates neural circuitry for threat detection. J. Cogn. Neurosci. **21**(8), 1523–1535 (2009)
17. Lindström, L., Kotiaho, J.S.: Signalling and reception. e LS (2001)
18. Lipkus, I.M., Hollands, J.G.: The visual communication of risk. JNCI Monographs **1999**(25), 149–163 (1999)
19. Moghimi, S., Kushki, A., Guerguerian, A.M., Chau, T.: Characterizing emotional response to music in the prefrontal cortex using near infrared spectroscopy. Neurosci. Lett. **525**(1), 7–11 (2012)
20. Öhman, A.: The role of the amygdala in human fear: automatic detection of threat. Psychoneuroendocrinology **30**(10), 953–958 (2005)
21. Pinsky, M., Sommer, L.: Pollution pods: can art change people's perception of climate change and air pollution? Field Actions Sci. Rep. J. Field Actions (Special Issue 21), 90–95 (2020)
22. Pourtois, G., Grandjean, D., Sander, D., Vuilleumier, P.: Electrophysiological correlates of rapid spatial orienting towards fearful faces. Cereb. Cortex **14**(6), 619–633 (2004)
23. Pourtois, G., Schettino, A., Vuilleumier, P.: Brain mechanisms for emotional influences on perception and attention: what is magic and what is not. Biol. Psychol. **92**(3), 492–512 (2013)
24. Silic, M., Cyr, D., Back, A., Holzer, A.: Effects of color appeal, perceived risk and culture on user's decision in presence of warning banner message. In: Silic, M., Cyr, D., Back, A., Holzer, A. (eds.) Effects of Color Appeal, Perceived Risk and Culture on User's Decision in Presence of Warning Banner Message. In Proceedings of the 50th Hawaii International Conference on System Sciences (2017)
25. Sommer, L.K., Swim, J.K., Keller, A., Klöckner, C.A.: "pollution pods": the merging of art and psychology to engage the public in climate change. Glob. Environ. Change **59**, 101992 (2019)
26. Spence, C., Ho, C.: Tactile and multisensory spatial warning signals for drivers. IEEE Trans. Haptics **1**(2), 121–129 (2008)
27. Vohra, K., Vodonos, A., Schwartz, J., Marais, E.A., Sulprizio, M.P., Mickley, L.J.: Global mortality from outdoor fine particle pollution generated by fossil fuel combustion: results from GEOS-chem. Environ. Res. **195**, 110754 (2021)
28. Zikmund-Fisher, B.J., et al.: Blocks, ovals, or people? Icon type affects risk perceptions and recall of pictographs. Med. Decis. Making **34**(4), 443–453 (2014)

Google Translate Facilitates Conference Abstracts' Acceptance, But Not Invitations to Deliver an Oral Presentation

Piotr Toczyski[1]([✉]), Grzegorz Banerski[2], Cezary Biele[2], Jarosław Kowalski[2], and Michał B. Paradowski[3]

[1] The Maria Grzegorzewska University, Warsaw, Poland
ptoczyski@aps.edu.pl
[2] National Information Processing Institute, Warsaw, Poland
[3] University of Warsaw, Warsaw, Poland

Abstract. Removing the language barrier could bring great benefits not only to the scientific community. Therefore, it is necessary to strive to improve both the tools and procedures in which these tools are used, to ensure a reliable exchange of knowledge. The authors try to find out whether the existing and widely available technology (Google Translate) contributes to the facilitation of knowledge sharing among scientists. Humanity has been trying to construct and improve the technology of universal real-time translation for a long time. For many, it was inspired by scifi works, in which, probably, this idea appeared already in the 1940s (see Leinster's "First contact"). This is an important topic because the language of science has long since become English, and for most of the scientific community it is not the mother tongue. Furthermore, we are now talking about the English languages of the world, or "world Englishes", not to mention those who say "the language of science is bad English". The paper tells a story which on the one hand constitutes a thoughtful anecdote, on the other may offer a good introduction to a serious scientific study. As it stands now, the main argument for including it is the story itself, with which we encourage further studies to scale our ideas in terms of a broader sample and comparability.

1 Background

Although Chinese researchers had reported their findings on the H5N1 avian influenza already in January 2004, the World Health Organization only found out about it after an international symposium several months later. Simply because the original paper was written in Chinese [21]. Would it look different today, with the ready availability of machine translation services such as Google Translate, Microsoft Translator, or DeepL at the fingertips of most Internet users? Since its launch in April 2006, the former online resource for quick translations offering 104 languages is now reportedly being used by more than 500 m users daily. What began as statistical machine translation has, since 2016, switched to neural machine translation using deep learning algorithms [15].

The new technology of machine translation has likewise been spreading among smartphone users, helping order food, ask for directions, talk about the weather, and understand classroom content and engage in meaningful interaction with the teacher. At the same time, widely available instructional videos are accompanied by jokes about the quality of translations [1], ridiculing cases of over-reliance on and uncritical use of technology.

For a long time, machine translation would have limited utility where language served more far-reaching functions than in a simple exchange, involving important nuances such as discussing personal values and ethical concerns or resolving conflicts. However, with Google's research on and development of the translation engine since 2016, which has been covered even in the popular media, users have been able to notice an improvement in the quality of the translations with the naked eye. The current opinion of many users from different business environments is that Google Translate provides by and large highly effective translations.

Here we investigate whether machine translation can make research more widely known and help researchers become part of the international academic community.

2 Literature on Google Translate

Deep learning improved Google Translate through the use of artificial neural networks [7]. Current literature on the tool's applications includes for instance a contrastive analysis of MT of Arabic verb forms and aspect into English [2], a comparison of the output quality of Google and Bing translators in Chinese-English translation [5], an evaluation of machine translation in more academic fields [14], a survey carried out among medical researchers admitting to using Google Translate to retrieve key study characteristics from the global literature [20], or a meta-analysis of the prevalence and incidence of traumatic tooth injuries which examined articles published in languages other than English [11].

There has also been a reflection on the correctness of translations from the perspective of preventing discrimination [22], and the reliance on Google Translate by students and universities [18]. Assertions have also been made that deployment of Google Translate may help promote a more inclusive international student environment [16], although some researchers express surprise at participants' readiness to use the tool to submit their assignments and responses [19].

To add to the picture, we check how Google Translate works in situations demanding a high level of communication, i.e., in scientific communication – a field dominated by the English language, but where most stakeholders are not its native speakers, and still not all can communicate in it. A language barrier is not key, but is often mentioned by respondents in quite recent studies, and may undermine the intention to even submit an application to a conference and participate in the international exchange of thoughts [4].

3 Objectives

We check whether universally available free translation technology can help achieve goals in an effective and constructive way in conditions close to natural conditions. Without the intention to understand and assess the underlying R&D behind MT, our approach is user-centered. We are interested in the end-user and the extent to which MT satisfies their need to effectively communicate an important issue requiring expertise.

Although roughly 500 million people use Google Translate for private purposes, we decided to tighten the acceptable translation standards. We used this tool to translate abstracts for submission to a conference in the field of social sciences focused on research conducted in Central and Eastern Europe.

4   Method

We carried out a natural experiment, without a control group in the design. Although comparable designs are the norm in psychological sciences, it is not necessary in a user-centered feasibility and applicability test of technology.

We have chosen a prestigious international conference at which the submitted abstracts undergo a double-blind review procedure. The assessment of abstracts is carried out by English-speaking experts in the relevant field.

Over the past four years, the conference had been receiving upwards of 1,900 submissions annually, at an acceptance rate ranging from 34% to 50%. The event boasts an established reputation; the scientific society organizing it had been founded over 50 years earlier and today boasts over 7 thousand members. Its mission is to understand people for the general good of humanity. The conference is not only a central event of the society, but is also considered by the organizers to be the most important international event in the scientific discipline concerned. Annually, the conference is attended by more than 3.5 thousand participants from various academic and practice-oriented sectors.

4.1   Choice of Academic Discipline

We have chosen the world of science, as its mission is to produce and develop knowledge. For scientists, language is an indispensable tool. What is more, according to the standards adopted in science, international circulation of one's output is usually a precondition for gaining tenure and recognition in a given field.

The world of science is conventionally divided into several fields, often imagined as distributed along a continuum space. On the axis between strictly technical sciences and humanities one can distinguish social sciences, often treated as an intermediate category. This is the category chosen in our study. It typically requires more verbose skills than mathematical notation and a higher level of linguistic nuance.

4.2   Selection of Abstracts

We used two sources of materials for research, external and internal. The external source (other conference submissions) served only to determine the optimal number of abstracts that would be suitable for submission to reflect the conditions close to natural. From the external source we managed to obtain 57 abstracts. These were approved presentations for the largest conference organized by a local Eastern European association of the relevant social science field in 2018. However, the limit for abstract submissions of 1,200 characters with spaces imposed by the organizer of the international conference we aimed at reduced our database of abstracts to just eleven. After machine translation from Polish into English by Google Translate the character limit of 1,200 characters was exceeded by two abstracts. Based on the procedure we could finally approve 9 potential abstracts for submission. However, we decided not to go forward, given the need to obtain copyright from the authors. For practical and ethical reasons, we were unable to obtain research material from this source. Possible complications of obtaining permission from individual authors to use the abstract and the expected time to obtain these permissions made the source inadequate. Instead, we collected the final research material from an internal source (our colleagues, researchers whom we knew personally). We compiled a database of our colleagues' Polish-language abstracts from the field of social sciences (internal source), which in previous years had been accepted for other (local) conference presentations, and which met the upper limit of 1,200 characters after translation into English with the help of Google Translate ($n = 18$). Finally, from this database we randomly selected 9 abstracts, which we used as research material.

These automatically translated texts contained grammatical errors, which we intentionally left uncorrected. The materials were supposed to reflect the situation of a researcher from a non-English-speaking country who intends to share their research with the international scientific community, uses only automatic translation, and does not correct the translation.

In this paragraph we present an example of such automatic translation (1 of 9) from the original Polish text into English, together with the title:

"Do viewers remember what irritates them?"

"Remembering advertising messages is one of the important parameters for assessing the effectiveness of advertising. The level of memorization of individual elements of the message is different and depends on many factors, such as the order in which they are presented, the total number of stimuli in the message, the personal importance of information, the clarity of information and their attractiveness to the viewer. In the study, which will be presented, respondents ($n = 1000$) in three independent, independent groups watched a properly prepared fragment of a real television program in which advertising messages (auto-promotional) differing in the way of assembly. The obtained results indicate the occurrence of sequence effects, the lack of influence of the number of stimuli, the positive effect of vividness of the message and the correlation between remembering and liking the message. The test results give guidelines for the design of these forms of promotion in the future."

As mentioned above, we used our own resources. The above conference abstract has been translated and reproduced from an unpublished source with the permission of the copyright holder who is one of the co-authors of the current paper.

4.3   Abstract Submission Procedure

The abstracts were submitted to the conference via e-mail accounts on a globally recognized portal. The accounts were registered as fictitious people with names matching different cultures, such as Zenon Kowalski, Felicia Williams, Carrie Cholmondeley, Mary Surren, etc. The applications were formally sent by academic teachers who had obtained their doctorates before 1 January 2016. The application form also required the submission of affiliations. We used the Academic Ranking of World Universities, from which we randomly selected nine universities in the middle of the ranking, located between the 400th and 500th places (e.g. East China Normal University, Bangor University, Federal University of Minas Gerais, etc.). The proportion of male and female names was 4:5. The nationality of the abstract submitter was determined by the country in which the university was located. From Wikipedia, we selected the most typical names for a given region of the world.

The research described in all abstracts was empirical, and these were submitted as original research intended for oral presentations. However, we expressed our willingness to participate in a poster session in case an abstract would not be approved by the reviewers for an oral presentation.

4.4   Ethics and Research Integrity

This leads us to the research integrity issue: whether program chairs have been previously asked as to whether they agree with this scientific experiment and whether the conference organizers were informed about any activities that were going to take place in the framework of the experiment. They were not, as this would invalidate all our research activity in this study. Our testing method is simple, has some originality and is well thought out in terms of not affecting the review process, and with a small chance for being discovered by the organizers. For ecological validity of natural experiments, the method of covert observation is justified as long as it is not harmful. As has recently been noticed, according to research integrity committees the consent from the participants in such scenarios is not necessary, as long as we focus on the positive consequences of the research, because if the social benefits of the research outweigh the cost, deception is acceptable [13]. Moreover, covert research is acceptable in some contexts, on condition that the researcher constantly questions the ethicality of their action and research, and its consequences [13]. In many very valuable interdisciplinary studies the researchers did not have to ask for the permission of the organization as otherwise the gatekeepers would have likely made the research difficult [13] and the behaviors to be observed would not have been visible to an overt observer [13]. To sum up: to uncover the reality of institutions, one does not ask them for consent. This is exactly the approach which we adopted in this paper.

We also analyzed the American Psychological Association's (APA) Ethical Principles of Psychologists and Code of Conduct, used globally as a framework of reference [3]. The APA Ethics Code clearly states in its Section 8 (Research and Publication, chapter 8.05 on "Dispensing with Informed Consent for Research") that: "Psychologists may dispense with informed consent only: "(1) where research would not reasonably be assumed to create distress or harm [AND] involves (a) the study of normal educational practices, curricula, or classroom management methods conducted in educational settings; (b) only anonymous questionnaires, naturalistic observations, or archival research for which disclosure of responses would not place participants at risk of criminal or civil liability or damage their financial standing, employability, or reputation, and confidentiality is protected; [or] (c) the study of factors related to job or organization effectiveness conducted in organizational settings for which there is no risk to participants' employability, and confidentiality is protected [OR] (2) where otherwise permitted by law or federal or institutional regulations" [3].

In our study we fulfill the above mentioned APA Ethics Code criteria: (Ad 1) The research did not reasonably envisage causing distress or harm; (Ad b) The research involved only naturalistic observations (which we mentioned as a natural experiment, with no control group). The disclosure of responses would not place participants at risk of criminal or civil liability or damage their financial standing, employability, or reputation. The confidentiality is fully protected. (Ad c) The study concerned factors related to academic job and academic organization effectiveness conducted in academic organizational settings for which there is no risk to participants' employability. The confidentiality is fully protected.

We did not wish to flood the conference organizers with hundreds or dozens of submissions, because to carry out a Google Translate feasibility only a few texts sufficed. In the submission process we abided by ethics principles. First of all, we did not add on much work for the reviewers, given that the abstracts never exceeded 1,200 characters. We also did not troll reviewers and organizers by sending control "lorem ipsum" or otherwise meaningless texts. These would not only pose an unnecessary workload and frustration to the scientific committee, but also not help to address the research question at hand.

The abstracts were veritable and relevant texts that had already been reviewed or accepted for print in the local scientific community. We had only collected texts of already proven scientific quality in the field of social sciences, and merely tested their potential for internationalization: we checked whether they would be seen as useful to the international scientific community when machine translated. We have not misrepresented scientific claims. The only distortion was the identity of the authors. However, we did not violate the rights of the authors, because we only collected our own texts and those by our colleagues.

We did not disrupt the logistics of the conference, because we only submitted a handful of abstracts. Therefore, this did not upset the evaluation of around 2,000 other applications. The principle of the conference is to consider the lack of post-decision reaction on the part of the author of the accepted work as a withdrawal of participation (e.g., for random reasons). This minimized the need for contacting and debriefing the organizers.

In effect the limitation of this study is the lack of a control condition. Although, as one of our reviewers notes, generating stimuli that sound meaningful, are linguistically and grammatically correct, but do not make sense when it comes to conveying scientific content would be feasible, it would be ethically questionable. We leave it to the next generations of authors to decide whether designing such a study would be appropriate from the research integrity viewpoint.

4.5 Results

Out of the nine abstracts submitted, in September 2018 eight were accepted for the conference – but only for poster sessions. One abstract was rejected. Table 1 presents an excerpt from the positive and negative evaluations sent out by the conference organizer in response to the submissions. The positive answer concerns the acceptance of the abstract only for a poster session. Table 2 shows the titles of the abstracts together with the decision whether or not to accept the submission in question.

Table 1. Positive and negative response of the conference organizer to the submitted abstract

Positive decision (abstract accepted for the poster session)	Negative decision (rejection)
Dear [...] Congratulations! [...] Our committee thought that the research described in your submission was information that [...] attendees will want to know, and we are excited to have it on the program	Dear [...] Thank you very much for your submission [...] Because the number of submissions was so high and space so limited, the committee was unable to include your presentation [...]

Table 2. Abstract titles and status of the proposed abstract

	Title (as translated by Google Translate)	Accepted as a presentation?	Accepted as a poster?
1	Do viewers remember what irritates them?	No	Yes
2	The impact of the program context and loyalty to the program on the effectiveness of advertising	No	Yes
3	Influence of shelf width on price perception and market share of sweets brands	No	Yes
4	Perception of the euro in Poland - economic and psychological factors	No	Yes
5	Emotions and visual attention - the influence of affect on the perception of art	No	Yes
6	The dynamics of visual attention and cognitive Engagement while reading hypertext	No	*No*
7	The cognitive and emotional consequences of extended interaction with the computer using eyesight	No	Yes
8	Patterns of visual attention in the process of solving textual math problems	No	Yes
9	The role of social support in preparing children and adolescents to deal with cyberaggression	No	Yes

4.6 Discussion of the Results

Almost all the machine-translated conference submissions were accepted, but only as posters.

We assume that if we had sent empty, randomly generated or "lorem ipsum" texts, the organizers would not have accepted them (we did not check this assumption for ethical reasons). The fact that none of the authors were invited to present a paper means that either the topic, its presentation, or the linguistic competence of the submitter was considered to be inadequate. In such a situation the poster format is a way for the organizers to nonetheless ensure the presence of many different ideas.

One abstract, however, did not go through the selection process. Why did this one not pass? A possible reason is that it was less consistent with the conference topic; it concerned cognitive rather than social processes, referring to an experiment using eyetracking and the analysis of oculographic indices, such as fixation length, saccade length and pupil dilation, thus being better suited to a cognitive rather than a social science venue.

One reviewer points out lack of information on the reviewers' language backgrounds and level of proficiency in English. While this is the case, we do not think it detracts from the implications of this study or is a limitation: in most scientific conferences, the identities of the reviewers are only known to the event chairs, and international events typically likewise recruit their scientific committee members and ad hoc reviewers from an international pool.

5 Implications

Since Google Translate is able to provide translation of abstracts at a level acceptable to the scientific community in the area of conference submissions, it means that a message translated in this way, including relatively complex ideas, concepts or problems, is understandable to the receiver. Thus, this technology can already be used in a wider formal context. Below, we identify possible applications of Google Translate and other analogous resources in formal cases.

5.1 Transfer of Scientific Ideas and Research Results

It is already possible to transfer knowledge between scientists from all over the world, also with the participation of researchers without language skills. These researchers do not have to be worse in their narrow disciplines than English- or other dominant-language-speaking researchers, and one should remember that the vast majority of today's scientific (and other) publications in English are penned by non-native users of this language [8, 10, 11]. The implementation of Google Translate into conference reporting systems could make the world's science more readily accessible and transparent, extend its outreach, and balance the distribution of scientific thought centers in specific fields.

5.2   Communication of Officials with Citizens and Non-citizens

Communication with offices and institutions could be made more efficient. Economic migrants, refugees, and other foreigners could participate in automatically translated interviews with officials, for instance via chat rooms, and automatic translation of official forms into the language of the applicant could be facilitated in a similar way as the translation of the content of websites, at least in the preliminary, provisional stages. Residence and work permits, social security, or opening a bank account are the basic formalities one needs to deal with during the first and most difficult period of one's stay in a new country. At a later stage it could also be possible to browse and search for job offers, prepare a CV and contact a prospective employer.

5.3   Wellbeing and Social Inclusion

We expect that in the long run, thanks to speech recognition and the intensification of communication in foreign languages in real time, translation technology can provide people with numerous personal benefits. We propound that the ability to express complex content can be important for one's image and foster success by enabling communication with diminished risk of face loss. Increasing communication is also important from the perspective of social inclusion, especially for vulnerable groups.

Recently the potential of voice assistant technology usage as a proxy for production of silver content (ie. creative, productive, wise and autonomous activity of older adults in new media) has been noted, together with number of new challenges it generates [17]. We noticed the interlingual potential of such technology using the translators mentioned in this article. Interactive and intelligent technology will be a substitute for social actors, preventing interlingual and international exclusion and disengagement. Voice assistants can substitute social interaction in a very restricted manner, but with the development of this technology, these interactions can become more meaningful and may become a way to provide aging people with the opportunity to maintain social interactions on the level necessary to stay active, even invisibly from a technological viewpoint [6], and at on the international level, should it be their choice.

5.4   Potential Threats

Among the consequences that we consider to be potentially socially negative, we discern a reduction in the motivation to learn foreign languages [7], but also an easier implantation of ideologies among young people, against which society will remain helpless. The disappearance of the language barrier opens up new spaces for propaganda, recruitment and attitude shaping that may undermine the social or legal order.

As is often the case, it seems that global and international science exemplifies the directions of other possible global social processes. While a facilitated exchange of scientific ideas and a removal of communication barriers may be of great benefit to the whole of humanity, and to the scientific community in particular, it is also worthwhile to carry out an analysis of opportunities and risks associated with machine translation.

According to our reviewers and other readers whom we consulted, even the current paper could make the reader suspect that it was translated using Google Translate, and therefore would place them in the role of participants in yet another study. But, as a reviewer writes: "On the other hand, isn't that really the point? Who can certify that this review has not been translated by Google Translate? And is it wrong?"

References

1. Abadi, M.: 4 times Google Translate totally dropped the ball (2017). https://www.businessi nsider.com/google-translate-fails-2017-11
2. Alasmari, J., Watson, J., Atwell, E.: A comparative analysis of verb tense and aspect in Arabic and English using Google Translate. Int. J. Islamic. Appl. Comput. Sci. Technol. **5**(3), 9–14 (2017) http://www.sign-ific-ance.co.uk/index.php/IJASAT/article/view/1683
3. APA: Ethical Principles of Psychologists and Code of Conduct. Including 2010 and 2016 Amendments. Effective date 1 June 2003 with amendments effective 1 June 2010 and 1 January 2017 (2016)
4. Banerski, G., Bohdanowicz, Z., Knapińska, A., Kopacz, A., Muller, A.: Remote Conferencing in the World of Science. National Information Processing Institute, Warsaw (2021). https://radon.nauka.gov.pl/analizy/zdalne
5. Chen, D.: A Linguistic Evaluation of the Output Quality of 'Google Translate' and 'Bing Translator' in Chinese-English Translation. Master thesis (2017)
6. Kowalski, J., Toczyski, P., Biele, C., Zdrodowska, A.: Reading is vital, but will it be Invisible? Screens vs. paper on our way to the naturalized technology of reading. In: Ganzha, M., Maciaszek, L., Paprzycki, M. (eds.) Annals of Computer Science and Information Systems, Volume 16. Position Papers of the 2018 Federated Conference on Computer Science and Information Systems. Polish Information Processing Society (2018). https://www.researchg ate.net/publication/327893548_Reading_is_Vital_but_will_it_be_Invisible_Screens_vs_P aper_on_Our_Way_to_Naturalized_Technology_of_Reading
7. Monroe, D.: Deep learning takes on translation. Commun. ACM **60**(6), 12–14 (2017)
8. Paradowski, M.B.: Review of the book by B Seidlhofer, Understanding English as a Lingua Franca: A Complete Introduction to the Theoretical Nature and Practical Implications of English used as a Lingua Franca. The Interpreter and Translator Trainer 7.2. [Special Issue: English as a Lingua Franca. Implications for Translator and Interpreter Education], pp. 312–320 (2013).https://doi.org/10.1080/13556509.2013.10798856
9. Paradowski, M.B.: Holes in SOLEs: re-examining the role of EdTech and 'minimally invasive education' in foreign language learning and teaching. English Lingua J. **1**(1), 37–60 (2015)
10. Paradowski, M.B.: What's cooking in English culinary texts? Insights from genre corpora for cookbook and menu writers and translators. Translator **24**(1), 50–69 (2018). https://doi.org/10.1080/13556509.2016.1271735
11. Pawlas, E., Paradowski, M.B.: Misunderstandings in communicating in English as a lingua franca: causes, prevention, and remediation strategies. In: Koutny, I., Stria, I., Farris, M., (eds.) Role of Languages in Intercultural Communication/Rolo de lingvoj en interkultura komunikado/Rola języków w komunikacji międzykulturowej, pp. 101–122, Poznań: Rys (2020). https://doi.org/10.48226/978-83-66666-28-3
12. Petti, S., Glendor, U., Andersson, L.: World traumatic dental injury prevalence and incidence, a meta-analysis - one billion living people have had traumatic dental injuries. Dent. Traumatol. **34**(2), 71–86 (2018). https://doi.org/10.1111/edt.12389
13. Roulet, T.J., Gill, M.J., Stenger, S., Gill, D.J.: Reconsidering the value of covert research: the role of ambiguous consent in participant observation. Organ. Res. Methods **20**(3), 487–517 (2017). https://doi.org/10.1177/1094428117698745 or http://eprints.nottingham.ac. uk/41131/1/Roulet%20et%20al%20ORM%202017.pdf or https://journals.sagepub.com/doi/ abs/10.1177/1094428117698745

14. Saffari, M., Sajjadi, S.: Evaluation of machine translation (Google Translate vs. Bing Translator) from English into Persian across academic fields. Mod. J. Lang. Teach. Methods **7**(8), 429–442 (2017)
15. Sommerlad, J.: Google Translate: How does the search giant's multilingual interpreter actually work? Independent (2018). https://www.independent.co.uk/life-style/gadgets-and-tech/news/google-translate-how-work-foreign-languages-interpreter-app-search-engine-a84 06131.html
16. Taylor, Z.W.: Intelligibility is equity: can international students read undergraduate admissions materials? High. Educ. Q. **72**(2), 160–169 (2018)
17. Toczyski, P., Kowalski, J., Biele, C.: Proxy users enable older people creative writing on the Web. Front. Sociol. **4**, 15 (2019). https://www.frontiersin.org/articles/10.3389/fsoc.2019. 00015/full
18. Todd, R.W.: An opaque engineering word list: which words should a teacher focus on? Engl. Specif. Purp. **45**, 31–39 (2017)
19. Van Miltenburg, E., Elliott, D., Vossen, P.: Cross-linguistic differences and similarities in image descriptions. In: Proceedings of the 27th International Conference on Computational Linguistics, pp. 1730–1741 (2017)
20. Versteegden, L.R.M., de Jonge, P.K.J.D., et al.: Tissue engineering of the urethra: a systematic review and meta-analysis of preclinical and clinical studies. Eur. Urol. **72**(4), 594–606 (2017)
21. Villar, R.: The importance of language. J. Hip Preserv. Surg. **5**(1), 1–2 (2018). https://doi.org/ 10.1093/jhps/hny002
22. Zou, J., Schiebinger, L.: AI can be sexist and racist – it's time to make it fair. Nature **559**(7714), 324–326 (2018). https://doi.org/10.1038/d41586-018-05707-8

Learning Affective Responses Through Evaluative Conditioning - New Developments in Affective Computing

Robert Balas[(✉)] and Grzegorz Pochwatko

Institute of Psychology, Polish Academy of Sciences, Warsaw, Poland
rbalas@psych.pan.pl
https://psych.pan.pl

Abstract. In affective computing, the system should accurately read the user's emotional states and respond to them adequately. It can also influence the user while having information about his temporary affective state. A vital psychological mechanism that refers to changes in affective responses due to simple pairings of stimuli is the evaluative conditioning (EC). Such form of learning might be of great importance in human-machine interactions, since such interactions are becoming a vital element of social relations that are construed, maintained, and ended via various artificial systems. In our studies, we investigated how physiological arousal moderates affective learning. Previous research showed arousal to play an important role in evaluative conditioning. However, those were declarative measures of self-assessed arousal levels. We used physiological measures of arousal and affect to show only moderate levels of arousal might support learning of affective responses. We discuss how our findings translate to the influence of artificial tools on social relations and attitudes. We also try to indicate a significant direction in the development of affective computing.

Keywords: Attitudes · Evaluative conditioning · Human-machine interaction · Social relations

1 Introduction

Ever since Picard [12] coined the term "affective computing", machines have gained great opportunities to monitor and analyze user reactions and respond adequately. Thanks to advanced technology, various systems are capable of detecting and recognizing the emotional states of users and moreover, they can try to influence them. Potential applications in the field of HCI are extremely wide: from serious, such as enriching communication in remote education, social robots in health care, safety systems at cars, to entertainment, e.g. interactive computer games or immersive storytelling in interactive VR experiences. One potential application is attitude formation. One can imagine, for example, an automatic training system that helps to shape proper eating habits or a virtual

© The Author(s) 2022
C. Biele et al. (Eds.): MIDI 2021, LNNS 440, pp. 259–266, 2022.
https://doi.org/10.1007/978-3-031-11432-8_26

environment for working with children with ADHD. A system equipped with the possibility of monitoring the level of arousal (even with the use of relatively simple measures, e.g. galvanic skin response - GSR) could use the evaluative conditioning effect to shape user's attitudes.

2   Theoretical Context

Carberry and de Rosis [2] described four main areas of affective computing development. They are: analysis and characterization of affective states, especially those exhibited in natural interactions, and analyses of the relationships between cognitive and affective processes (e.g. in learning environments); automatic recognition of affective states (e.g. facial expressions, sounds, or physiological responses); adapting system response to affective states of a user; last but not least, designing affective virtual agents (virtual humans that are able to use the information on user state to exhibit proper affective answer, or to influence user's state in enhanced HCI). Affective computing methods have been used in many fields, including for example training and learning environments. As emotions have impact on motivation, reasoning or decision making it is feasible and necessary to also look at the consequences of eliciting affective states by the elements of interface (like conversation agents, virtual humans), not only detecting and responding to (see e.g. [9]). With the above in mind, we would like to check to what extent interface users can be influenced by affective information, for example in a conditioning procedure. More precisely, we want to trace the process of forming attitudes in a situation where the user encounters emotional stimuli, the source of which is a computer system.

The purpose of the current studies was to determine the dynamics of emotional processes in attitude formation in evaluative conditioning (EC). EC is defined here as a change in the evaluation of a specific stimulus (CS - conditioned stimulus) that can be attributed to its repeated presentation with another affective stimulus (US - unconditioned stimulus; cf. [6,7]). In three experimental studies (N = 94) we investigated the relationship between the level of arousal generated by the US and the effectiveness of evaluative conditioning, the extinction of this effect, and the effectiveness of revaluation of the affective stimulus. Previous research has suggested an effective transfer of the arousal value of the affective stimulus US to the conditioned stimulus CS, vaguely suggesting that the evaluative conditioning does not only involve declarative ratings of liking, but also occurs at the arousal level [4]. However, to this point, only one paper investigated this hypothesis analyzing only the declarative arousal ratings of subjects. Key aspects of this project were (a) to trace how arousal affects the size of the effect of EC, (b) to determine possible asymmetries between positive conditioning (US+) and negative conditioning (US−), (c) to determine the effect of the arousal value of the US stimulus on the ability to extinguish the conditioned evaluative response or change its sign, and (d) to determine the degree of intentional control over the acquisition and expression of conditioned affective responses depending on the arousal value of emotional stimuli [1,3,5]. It is worth emphasizing that in our case the computer system receives information about the user's physiological arousal and uses it in real time to adjust the stimulation.

3   Experiments

3.1   Method

Participants

Ninety-four participants volunteered to participate in the experiment (46 Female) ranging in age from 18 to 35 years ($M = 24.8$, $SD = 4.21$). Participants were individually tested.

Materials and Procedure

All three experiments used neutral human faces as CSs and affective images differing in sign (positive vs. negative) and arousal level (low vs. high) as USs. CSs were selected from a validated set of facial expressions [11] and USs from IAPS [8]. The latter were pre-selected based on available standardization data to maximize differences on two orthogonal dimensions: affective value (positive vs. negative) and arousal level attributed to the images (low vs. high). Therefore, we managed to create four groups of USs used in the studies: positive – highly arousing, positive – little arousing, negative – highly arousing, and negative – little arousing.

All subsequent tests were conducted in a standard EC procedure with two stages at its core. After filling in a consent form, participants were instructed to observe CS-US pairs appearing several times in random order (presented together in Experiment 1 or in rapid succession in Experiments 2 and 3). We used eight different CS-US pairs. In the second stage, participants made evaluative judgments of the CS stimuli using a scale from 1 – very negative to 9 – very positive.

GSR and EMG were continuously measured during both stages of the study with a BIOPACK MP150 set. Galvanic skin response (GSR) refers to dynamic changes in sweat gland activity that correlate with the intensity of one's emotional state (i.e. emotional arousal). Electromyography (EMG) measures muscle response and was used in two facial muscles: the corrugator supercilli that is typically active when frowning (negative emotion), and the zygomaticus major that is active when lifting the corners of the lips when smiling (positive emotion). The EMG of both muscles is believed to reflect positivity and negativity of the emotional state of the participants.

Participants were thanked and debriefed after experiment completion.

In each study, within-group manipulations and random assignment of CS-US pairs were used. The description of subsequent experiments includes only those elements of the procedure that differed from the standard study design mentioned above.

Results

In the first (N = 31) and second (N = 29) studies, the goal was to determine the relationship of physiological arousal to the size of the effect of evaluative conditioning. Selected US stimuli with low and high arousal values were presented in pairs with neutral CS stimuli. We hypothesized that high-arousing stimuli would lead to a stronger evaluative conditioning effect compared to low-arousing US. Our results revealed an average EC effect of higher evaluative arousal for positively conditioned CSs compared to negatively conditioned ones, $F(1,59) = 4.54$, $p < .05$. Furthermore, declarative scores of arousal gathered after presentation of negatively conditioned CS were found to be higher than ratings after presentation of positively conditioned CS, and declarative arousal increased after conditioning, $F(1,59) = 6.29$, $p < .01$ (Fig. 1). Finally, the study showed that the level of physiological arousal after the presentation of the CS stimulus increased during the conditioning stage with increasing number of presentations of US-CS pairs and persisted throughout the postconditioning stimulus ratings. The results support the endorsed hypothesis that the magnitude of the conditioning effect varies with the arousal elicited by the US stimulus and that there is a transfer of arousal between the US and CS both at the declarative level and at the level of the physiological response that is a marker of arousal.

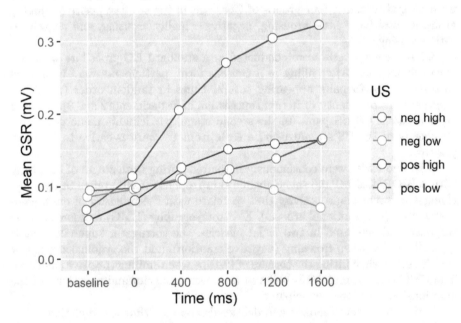

Fig. 1. Level of arousal in response to conditioned CS stimuli as a function of the affective value of the US (pos - positive, and neg- negative) and the level of arousal elicited by the US (high vs low).

In the third experiment (N = 34), the objective was to determine the mechanism responsible for the transfer of arousal between the affective stimulus US and the neutral stimulus CS. In this study, we manipulated the level of arousal of the affective stimuli (low and high arousal) and the order of presentation of the stimulus during the conditioning phase (CS-US or US-CS). The hypothesis tested was that physiological arousal transfer is more efficient when the affective stimulus US (causing arousal increase) is presented first, followed by the neutral stimulus CS, compared to the opposite order of stimulus presentation. We did not assume such a difference for the declarative measure of arousal because we assumed that transfer at the declarative level of the measure does not depend on the order of presentation of the CS and US stimuli. The results confirmed the hypothesis that the transfer of arousal at the declarative level does not depend on the order of presentation of the stimulus. On the contrary, we did not obtain the expected effect of greater arousal transfer at the physiological response level for US-CS pairs compared to CS-US pairs.

We also hypothesized that the level of arousal of conditioned CS stimuli would depend on the temporal interval between US and CS such that transfer would be greater for CS presentations at the peak of the arousal response to US compared with CS presentations earlier and later relative to the highest response to US. The results of this study confirmed the previously obtained effect of physiological arousal transfer between US and CS at both the declarative and physiological response levels. Furthermore, the transfer of arousal at the declarative rating level did not depend on the temporal interval between the US and CS. For physiological arousal elicited by conditioned CSs, a higher transfer was observed for those CSs presented at the peak of the US response (during conditioning). The latter effect was present only for negatively conditioned CSs, suggesting (as in previous studies) the existence of an asymmetry in responses to positive and negative stimuli.

4 Discussion

The project succeeded in accomplishing most of the stated substantive goals attributed to the primary research. Firstly, a novel line of research was conducted on the transfer of arousal (at the declarative level and at the level of physiological reactions) in evaluative conditioning, which is an important addition to the knowledge on the mechanisms of attitude acquisition and their dependence on stimulus characteristics. Second, we examined for the first time how the procedural characteristics of EC (CS-US presentation order) are related to the arousal value of affective stimuli. And third, new experimental procedures were developed to enable complex research on the role of arousal in attitude acquisition.

Let us consider the many possible applications of the above discoveries. Virtual environments in which the user communicates in a natural way are becoming more and more common. This applies not only to experimental laboratory environments, but also to commercial and entertainment applications. Both simple interfaces (like chatbots [10,14]) and more complex environments with some or many virtual humans (e.g. [13]) can be effective. More research is needed on this phenomenon.

5   Limitations

One of the basic limitations of the interpretation and application of the above-described results for a HCI and affective computing is the nature of the experimental procedure used. On the one hand, it allows for precise control of distorting variables, on the other, it does not resemble natural situations in which the system could try to influence the participant's affective arousal. It would benefit from moving it to a more realistic setting, e.g. during an interaction between participants using social media.

Another limitation is the use of precise equipment that records physiological responses incomparably more accurately than commercially available sensors. The use of arousal peaks detection algorithms developed by us requires uninterrupted registration with millisecond precision, which may turn out to be impossible in the case of e.g. commercial VR HMDs, game pads or watches.

6   Further Directions

In the times of the Internet of Things and the widespread use of measuring devices, affective information will gain more and more importance (both the knowledge about the affective state of the system user and the ability to influence it). Understanding the processes of evaluative conditioning and algorithms to effectively detect a user's affective states in real time based on the available information is essential for finding applications as well as for deepening the basic understanding of the subject. For example, modern consumer VR systems are equipped with sensors that capture large amounts of data for simulation purposes. This data can also be used to infer about affective arousal, and this information can then be used in turn to modify the content of the simulation. Subsequent research should focus on users in their natural environments, on the daily relationships of users and technology. If your headset knows about your emotions, why not use it to help you. It is especially important in times when we experience isolation and more and more of our daily relationships are mediated by information systems. Not to mention the replacement of relationships with humans with relationships with machines.

Acknowledgments. This project has received funding from the National Science Center OPUS grant no. 2015/17/B/HS6/04204.

Conflict of Interest. Both Authors declare no conflict of interest concerning the work presented in this paper.

Ethical Statement. Both Authors declare that the research reported in this paper is conducted in accordance with general ethical guidelines in psychology as outlined in European Commission Ethics in Social Science and Humanities (2018).

References

1. Balas, R., Gawronski, B.: On the intentional control of conditioned evaluative responses. Learn. Motiv. **43**(3), 89–98 (2012)
2. Carberry, S., de Rosis, F.: Introduction to special issue on 'affective modeling and adaptation'. User Model. User-Adap. Inter. **18**(1–2), 1–9 (2008). https://doi.org/10.1007/s11257-007-9044-7
3. Gawronski, B., Balas, R., Creighton, L.A.: Can the formation of conditioned attitudes be intentionally controlled? Pers. Soc. Psychol. Bull. **40**(4), 419–432 (2014)
4. Gawronski, B., Mitchell, D.G.: Simultaneous conditioning of valence and arousal. Cogn. Emot. **28**(4), 577–595 (2014)
5. Gawronski, B., Mitchell, D.G., Balas, R.: Is evaluative conditioning really uncontrollable? A comparative test of three emotion-focused strategies to prevent the acquisition of conditioned preferences. Emotion **15**(5), 556 (2015)
6. Hofmann, W., De Houwer, J., Perugini, M., Baeyens, F., Crombez, G.: Evaluative conditioning in humans: a meta-analysis. Psychol. Bull. **136**(3), 390 (2010)
7. Jones, C.R., Fazio, R.H., Olson, M.A.: Implicit misattribution as a mechanism underlying evaluative conditioning. J. Pers. Soc. Psychol. **96**(5), 933 (2009)
8. Lang, P.J., Bradley, M.M., Cuthbert, B.N.: International affective picture system (IAPS): affective ratings of pictures and instruction manual. Technical report A-8, University of Florida, Gainesville (2008)
9. Mejbri, N., Essalmi, F., Jemni, M., Alyoubi, B.A.: Trends in the use of affective computing in e-learning environments. Educ. Inf. Technol. **27**, 3867–3889 (2022). https://doi.org/10.1007/s10639-021-10769-9
10. Nadri, C., et al.: Emotion GaRage Vol. II: a workshop on affective in-vehicle display design. In: 12th International Conference on Automotive User Interfaces and Interactive Vehicular Applications, AutomotiveUI 2020, pp. 106–108. Association for Computing Machinery, New York (2020). https://doi.org/10.1145/3409251.3411736
11. Olszanowski, M., Pochwatko, G., Kuklinski, K., Scibor-Rylski, M., Lewinski, P., Ohme, R.K.: Warsaw set of emotional facial expression pictures: a validation study of facial display photographs. Front. Psychol. **5**, 1516 (2015)
12. Picard, R.W.: Affective computing. Technical report, M.I.T Media Laboratory Perceptual Computing Section Technical Report No. 321 (1995)
13. Volonte, M., Hsu, Y.C., Liu, K.Y., Mazer, J.P., Wong, S.K., Babu, S.V.: Effects of interacting with a crowd of emotional virtual humans on users' affective and non-verbal behaviors. In: 2020 IEEE Conference on Virtual Reality and 3D User Interfaces (VR), pp. 293–302 (2020). https://doi.org/10.1109/VR46266.2020.00049
14. Weber-Guskar, E.: How to feel about emotionalized artificial intelligence? When robot pets, holograms, and chatbots become affective partners. Ethics Inf. Technol. **23**, 601–610 (2021). https://doi.org/10.1007/s10676-021-09598-8

Attitudes Towards Online Social Interactions and Technology in the Offering of Help During the COVID-19 Pandemic

Gabriela Górska[1,2]([✉]), Oliwia Maciantowicz[3], Malgorzata Pawlak[4], and Olga Wojnarowska[1]

[1] National Information Processing Institute, Warsaw, Poland
gabriela.gorska@opi.org.pl
[2] Robert Zajonc Institute for Social Studies, Warsaw University, Warsaw, Poland
[3] Faculty of Psychology, Warsaw University, Warsaw, Poland
[4] Institute of Psychology, Polish Academy of Science, Warsaw, Poland

Abstract. As the COVID-19 pandemic confined millions across the globe to their homes, technology proved an indispensable tool that allowed humanity to sustain many aspects of everyday life, including social behaviours. In compliance with quarantine restrictions, communities were unable to support each other in the usual manner; simultaneously, the demand for such support grew, owing to the difficult circumstances. This study (N = 196) explores whether technology enabled or hindered this specific type of social interaction – helping others. We discovered that General Online Social Interaction Propensity correlated positively with helping – although it demonstrated stronger correlations with online support. The Technology Adoption Propensity Index *Optimism* and *Proficiency* sub-scales failed to correlate significantly with helping and only *Vulnerability* subscale showed significant correlation. In conclusion, both GOSIP and TAP *Vulnerability* are valid predictors of proneness to offer help online. We suggest considering various personality predispositions may help to maximize the effectiveness of online helping.

Keywords: COVID-19 pandemic · General online social interaction propensity · Prosocial behaviour · Technology adoption propensity

1 Introduction

During the 2020 COVID-19 pandemic, millions around the globe have found themselves to a large extent confined to their homes for weeks on end. Social distancing measures and lockdown regulations have suspended or limited a wide variety of activities that involve in-person interactions [1]. Yet, the need to help others has prevailed, and a host of online initiatives have launched via social media [2].

In Poland, numerous groups offering community help emerged. One of the fastest-growing support groups in Poland, 'Visible Hand' (*Widzialna Ręka*), amassed over 90,000 members during the first weeks of its existence [3].

© The Author(s) 2022
C. Biele et al. (Eds.): MIDI 2021, LNNS 440, pp. 267–276, 2022.
https://doi.org/10.1007/978-3-031-11432-8_27

1.1 Online Interaction

Despite online interaction's mirroring of 'real-life' behaviour, it retains a number of distinct characteristics – many of which have become subjects of research (for a review of the current state of knowledge, see [4]). Studies have demonstrated that online engagement varies, depending on the characteristics of online platforms [5, 6]. Another factor that plays a role in how people experience technology mediated interactions is the form these interactions take. A recent study on copresence and well-being during COVID-19 lockdown showed that copresence experienced by respondents was positively correlated with time spent on video calls (as opposed to audio-only calls) [7]. Importantly, personal disposition affects decisions to engage (or not) in online behaviours; individuals differ in their propensity to interact with others through the internet [8]. Other traits and possible factors identified as those that moderate human-technology engagement and adoption are psychological resilience, optimism, innovativeness, self-efficacy, habit, social influence, risk-taking [9], and individuals' perception of their security online [10, 11]. Considering the aforementioned research, it is transparent that multiple factors come into play in decisions on whether to engage with others online.

1.2 Help Mediated by Technology

A subset of online social interactions that is less widely researched is altruistic behaviours. The body of literature on the differentiation between online and offline helping behaviour, and on how technology mediates online prosociality, remains limited.

Several studies examining prosocial behaviour online suggest that helping others online is influenced by altruism and reciprocity [12–15]. Moreover, literature indicates that online and offline prosocial behaviour are positively related [12, 15]. Wright [14] demonstrated that patterns of cyber prosocial and antisocial behaviour imitate their real-life counterparts.

Some personality traits that supported the helping behaviour exhibited by its subjects differed depending on, for instance, their relation to the person who received the help. The participants' openness to experience contributed to their tendency to help only when strangers were the recipients [16]. Moreover, extraversion contributed to all helping behaviours, while agreeableness only enhanced their tendencies to help friends [16, 17].

Also, situational factors can affect altruism. Public acts of help bring promises of reciprocity, and attract so-called *egoists*, who are primarily interested in the personal benefits that result from helping. In contrast, *altruists* more often decide to act prosocially, regardless of personal gain [18].

In the case of helping behaviours, social media enables feelings of engagement with no more than a click, without subsequent action that would realistically fulfil others' needs (see *clicktivism*; [19, 20]). Although online donations reach wider networks, they do not necessarily lead to more frequent (or of higher monetary value) donations [21].

Considering the above, we suspect that specific characteristics exist that influence online interactions, such as willingness to interact with others and perception of technology.

The Technology Adoption Propensity (TAP) index and the General Online Social Interaction Propensity (GOSIP), scales rooted in the psychology of individual differences, can assist in predicting prosocial behaviour.

TAP aims to identify the personal variables that moderate the use of technology [11]. Its authors identified four subscales on which human-technology interaction relies: two are considered contributors (*Optimism* and *Proficiency*); and two are inhibitors (*Vulnerability* and *Dependence*). Their purpose is to indicate whether subjects perceive technology as enabling them to meet their goals more conveniently (*Optimism*); whether subjects consider themselves able to swiftly master new technologies (*Proficiency*); how strongly subjects believe that using technology exposes them to being exploited (*Vulnerability*); and how likely subjects are to become unhealthily dependent on technology (*Dependence*).

GOSIP scale was developed to understand how personality influences involvement in online conversations [8]. As with *willingness-to-communicate*, an individual trait responsible for the tendency to initiate and engage in face-to-face interactions [22], the authors of the GOSIP scale attempted to measure individuals' tendency to be involved in online communication [8]. The measure aims to encompass several aspects of online communication: their enjoyment of online communication; the social needs covered by online interactions; and the role of individuals in group dynamics [8]. These factors contribute to a general disposition towards online engagement that can mediate human-to-human online communication.

1.3 Current Study

We were unable to identify any studies that link technology adoption and propensity for online social interaction to prosocial behaviour; therefore, our hypotheses are driven by other effects described in the literature.

Since openness to experience and extraversion have previously been linked to prosocial behaviour, we expected that propensity for online social interaction would demonstrate similar effects, in addition to significant connections with higher frequency and higher engagement in helping activities (Hypothesis 1a for online and 1b for offline behaviours). We hypothesised that the TAP subscales would connect significantly with individuals' prosocial behaviour. We expected that TAP *Proficiency* and *Optimism* would correlate positively (Hypotheses 2a and 2b, respectively), and TAP *Vulnerability* negatively with participants' willingness to participate in helping activities (Hypothesis 2c).

Online and offline activities must be differentiated. We predicted that both TAP and GOSIP would exhibit stronger correlations with online than offline behaviours (Hypothesis 3a for TAP and 3b for GOSIP), since both constructs describe individual tendencies pertaining to technology and internet use. Finally, we hypothesised that TAP and GOSIP were independent predictors of individuals' helping behaviours (Hypothesis 4a for TAP and 4b for GOSIP), since social interaction and propensity for technology are clearly distinct in the context of the definitions [8, 11].

2 Method

2.1 Participants and Procedure

The study was conducted during the first Polish lockdown at the onset of the COVID-19 pandemic (March-May 2020). Participants were recruited via social media: invitations were published in Facebook groups comprising individuals requesting and offering community help, in addition to groups for university students. Any adult volunteer could participate. Informed consent was obtained from all participants. Aims of the study were not masked. Study was implemented in accordance with the 1964 Helsinki Declaration and its later amendments. Study was approved by the ethics committee of the National Information Processing Institute, Warsaw, Poland. Study was fully anonymous; participants' personal data was not collected. A preliminary study' report was available via email. The participants, totalling 234, completed questionnaires. One observation was excluded due to missing data. With consideration for the aims of the study, we analysed only the data from the participants who had declared at least one prosocial activity. Thirty-seven participants (21 women and 16 men, with a mean age of 31.4) declared no such activity; therefore, only 196 observations (162 women, 29 men, and 5 who declared 'other', with a mean age of 34) were further analysed.

2.2 Measures

Three independent sub-scales of the TAP index [11] were used to investigate how the characteristics they describe (Optimism, Proficiency, and Vulnerability) related to the frequency and extent to which subjects engaged in helping behaviours, both online and offline. The *Dependency* subscale was excluded since it was judged irrelevant to our hypotheses. A Polish language translation, courtesy of J. Kowalski, can be found in Appendix 1. In the original study, the internal consistency coefficients varied between 0.73 and 0.87 [11].

A Polish language translation of the GOSIP scale [8] was produced for the purpose of the study. It was generated based on four independent translations and evaluated by four independent experts. The final translation can be found in the Appendix 1. The internal consistency coefficient reported by the authors was 0.92 [8].

Engagement in helping behaviours was assessed using a set of questions devised for the purpose of this study (see Appendix 2). Subjects were asked to state whether they participated in various helping activities during the ongoing lockdown. The activities on which the questionnaire enquired were divided into two groups (online and offline), and included financial or psychological support, and organizing social events online. Responses were given on a scale describing the frequency of occurrence of each behaviour (from 'never' to 'daily'). Consequently, we were able to discern two dependent variables: the number of activities (the number of declared types of activity engaged in); and the frequency of helping (the accumulated frequencies of the declared behaviours). An additional variable, a number of activities x frequencies was created as a simple accumulation of the two previous variables, which reflects both factors.

The reliability coefficients (Cronbach's Alpha) for TAP and GOSIP are presented in Table 1.

3 Results

Table 1 presents the correlations, means, standard deviations, and internal consistency coefficients for each of the variables used in the study.

Statistical analyses were conducted using IBM SPSS Statistics 26 software. Hypothesis 1a was confirmed: propensity for online social interaction demonstrated significant positive correlation with the declaration of online behaviours (from 0.22 to 0.30; $p <$ 0.01 for the correlations with declarations of online behaviours – see Table 1). However, the correlation with offline behaviours was not confirmed (Hypothesis 1b; 0.5 to 0.10; p > .05). The FDR correction for multiple comparisons revealed significant results [23]. TAP, however, demonstrated significant effects only on the *Vulnerability* subscale (Hypothesis 2c), which correlated significantly and negatively with all categories of helping (from -0.27 to -0.17; $p < 0.05$ for all the correlations with online and offline helping declarations – see Table 1). All the correlations with TAP *Vulnerability* remained significant ($p < 0.05$) following the FDR correction for multiple comparisons [23]. However, Cronbach's Alpha was moderately low for TAP *Vulnerability* compared to other subscales (see Table 1; see more in the Limitations section).

The correlation with TAP *Optimism* (Hypothesis 2a) and TAP *Proficiency* (Hypothesis 2b) did not correlate significantly with helping; thus, only Hypothesis 2c was confirmed.

Table 1. Correlation coefficients, means, standard deviations and internal consistency coefficients.

	1.	2.	3.	4.	5.	6.	7.	8.	9.	10.
1. Frequency of occurrence: online	–									
2. Frequency of occurrence: offline	.37**	–								
3. Number of activities types: online	.89**	.47**	–							
4. Number of activities types: offline	.24**	.86**	.37**	–						
5. Number x frequency: online	.87**	.57**	.89**	.40**	–					
6. Number x frequency: offline	.35**	.86**	.53**	.76**	.64**	–				
7. TAP: optimism	−.10	.10	−.11	.08	−.05	.12	–			
8. TAP: proficiency	−.07	.03	−.08	.06	−02	.09	.37**	–		
9. TAP: *vulnerability*	−.18*	−.17*	−.27**	−.17*	−.22**	−.26**	.09	−.05	–	
10. GOSIP	.28**	.07	.30**	.09	.22**	.10	−.15*	−.16*	.05	–
M	3,76	3,59	1,90	1,88	12,30	11,99	2,05	2,48	2,14	3,92
SD	3,65	3,63	1,60	1,68	22,44	30,31	0,76	1,01	0,83	1,46
α	–	–	–	–	–	–	.81	.84	.63	.93

*$p < .05$; ** $p < .001$

To test whether TAP and GOSIP exhibited stronger relations with online than offline behaviours (H3a and H3b), we used the Fisher R-to-Z-transformation. For TAP (H3a), our analyses suggested that it was not more strongly connected with the type of declared helping activities, (Z = 0.126, p = 0.45 for online/offline frequency of occurrence; Z = 1.28, p = 0.10 for number of activities; and Z = −0.515, p = 0.303 for number x frequency). For GOSIP (H3b), the relationships with online frequency, number of activities and variable combining frequency, and number of activities were stronger than those declared offline (Z = 2.664 p = 0.004; Z = 2.678 p = 0.004; and Z = 1.998, p = 0.023, respectively).

Finally, to test whether TAP and GOSIP were independent predictors of helping behaviours (Hypotheses 4a and 4b), we conducted a series of regression analyses (see Table 2) with frequency and number of declared helping activities as dependent variables and TAP *Vulnerability* and GOSIP as predictors. The analyses revealed that TAP *Vulnerability* (H4a) and GOSIP (H4b) independently explained the variance of helping behaviours. Moreover, we observed that TAP *Vulnerability* alone was significant in explaining offline behaviour, whereas, during analysis of online activities, both TAP *Vulnerability* and GOSIP were significant predictors. GOSIP, however, had higher incremental R in all models that analysed online activities, adding more explained variance to each model when entered to the analyses after TAP *Vulnerability*. In summary, Hypotheses 4a and 4b were confirmed extent: GOSIP and TAP proved to be independent predictors of online helping – although only the TAP *Vulnerability* subscale was considered.

Table 2. Regression analyses with type of help as dependent variables with TAP *Vulnerability* subscale and GOSIP as predictors.

Predictors\Criterion	Frequency of occurrence: online		Frequency of occurrence: offline		Number of activities types: offline		Number of activities types: online		Number x frequency: online		Number x frequency: offline	
	β	ΔR^2	β	ΔR^2	B	ΔR^2	β	ΔR^2	β	ΔR^2	β	ΔR^2
TAP: *Vulnerability*	.17*	.03*	.17*	.03*	.16*	.03*	.25**	.06**	.21*	.04*	.26**	.07**
GOSIP	.27**	.07**	.06	.01	.08	.06	.25**	.08**	.21*	.05*	.09	.01

* p < .05; ** p < .01; ΔR = incremental R for each predictor when entered after the other predictor.

4 Discussion

This study investigated the relationships of technology adoption and propensity for general online social interaction with online and offline helping activities during the first Polish COVID-19 lockdown. The results demonstrated support for Hypothesis 1a: propensity for social interaction related significantly to higher frequency and to higher numbers of declared helping activities, both online and offline. In TAP, only the *Vulnerability* subscale correlated significantly with the dependent variables (Hypothesis 2c). The results returned small negative correlations with both the number of activities and the number of frequencies. In propensity for general online social interaction, Hypothesis 3b was also confirmed: the correlations were significantly higher for online than for offline behaviours. In TAP (H3a), the difference was insignificant: no stronger correlation for online activities was observed. For online activities, TAP and GOSIP were significant and independent predictors of offering help. TAP also proved a significant predictor for offline helping offers (Hypotheses 4a and 4b).

It is noteworthy that the majority of items in the questionnaire required high degrees of interaction with others. It was unsurprising, therefore, that the respondents' results on the GOSIP scale correlated positively with the frequency and the number of types of helping behaviour they offered. Predictably, this relationship was stronger for online activities, since GOSIP was designed to measure individuals' propensity to engage specifically in online interactions. Yet, it would be reckless to conclude that propensity to interact with others is a straightforward predictor of more general altruistic behaviours. As other study reports, the relationship between extraversion and altruism might not be linear, but rather U-shaped, meaning when playing the dictator game, individuals with very high or very low extraversion scores ($-2SD$) were more likely to give a higher percentage of the endowment to the other player than those who scored moderately low ($-1SD$) [17]. With this knowledge, charity groups or services would benefit from enabling various ways of engaging with their causes. The altruistic potential of highly extroverted individuals can easily be harnessed in direct forms of assistance, while introverts would more likely engage in other forms – specifically, volunteer work that can be accomplished alone, without the need to interact with or approach others.

The negative correlation between *Vulnerability* and the frequency of engagement in helping activities, both online and offline, compels us to enquire why this technology-related scale applies in a similar way to such behaviour, regardless of whether it demands the use of technology. An individual's score on the *Vulnerability* scale should reflect to what extent that person believes that using technology would increase their chances of falling prey to malicious schemes, or of becoming a victim of financial fraud or identity theft. These fears might be related to anxious personality disposition [24, 25], which, in turn, might inhibit individuals' proneness to empathetic reactions. There is evidence that experimentally evoked anxiety decreased the strength of empathy responses [27, 28]. Our results align with these findings, as we failed to prove the technology-specific aspect of TAP *Vulnerability* subscale. Anxiety might limit both online and offline altruistic behaviours similarly. It may inhibit one's initiative to act or react due to threatening interpretations of the reality, or concentration on one's own safety. Lam, Chiang and Parasuraman found that feelings of insecurity around technology use were rare for services deemed low-risk – even among individuals who tended to deem technology unsafe [28]. The designers of successful online platforms understand that the fear of being exploited via technology can be mitigated by a variety of social cues [29]. Users' perception of the extent to which they are capable of applying safeguarding strategies helps to predict whether they will decide to engage in helping behaviours or not.

Our findings demonstrate that although GOSIP is a significant predictor of online community assistance, its accuracy fails to extend offline. We conclude that GOSIP measures a technology-specific type of engagement in social interactions. This might also suggest that the social aspect of helping behaviours varies between the online and offline worlds. We may conclude that those who are socially engaged online could also be effective online helpers. TAP and GOSIP both independently predicted engagement in helping behaviours. Such results indicate that separate processes are responsible for the effects. The absence of significant correlation between TAP *Vulnerability* and GOSIP appears to confirm this interpretation. It is probable that individuals are differently motivated to help; thus, when facilitating helping behaviours, it is imperative that this diversity is accommodated. The independence of TAP and GOSIP in this study also offers evidence for differentiation between the two constructs, which to our knowledge, have not previously been tested. The lack of correlation between TAP *Vulnerability* and GOSSIP may appear surprising, since in the eyes of many a tendency to interact with others should be negatively connected with one's internet mistrust. However, in our opinion independence of GOSSIP and TAP *Vulnerability* simply adds to perceived accuracy of both constructs: GOSSIP reflects individual attitudes toward social online interactions and TAP towards technology.

4.1 Limitations and Future Directions

This study was correlational in nature and the participants were recruited on social media platforms. Conducting a similar study with the use of a more representative sample could serve to eliminate this limitation. An experimental approach would allow us to explore ways of facilitating helping behaviours by proposing different types of engagement that are better suited to individuals – including those who possess lower extraversion or need for security. Future studies could investigate whether other variables describing individual traits, such as extraversion or neuroticism, act as predictors of online altruism. Such research could potentially mediate the effects reported in this study.

Moreover, the progression of the COVID-19 pandemic could entail limited offline help due to the public anxiety around such exceptional circumstances. The number of suggested online activities, therefore, was likely inflated, when compared to the 'normal', pandemic-free context; thus, a future study with consideration for helping methods prior to the emergence of COVID-19 could prove useful. Consideration for the perceived pandemic threat could serve as a relevant mediator to such a study.

This study has also elucidated issues on the content validity of applied questionnaires. A study on the external validity of TAP would be compelling in terms of exploring usability –particularly that of the TAP *Vulnerability* subscale – which is not cyber-specific and reaches beyond the personality traits on which this study focused. Additionally, the Cronbach's alpha coefficient of our results attained by TAP was relatively low; it is noteworthy, however, that the authors of TAP reported similarly low alpha coefficients, which probably also results from the length of the scale, i.e., TAP *Vulnerability* subscale consists of only 3 items [11]. Last but not least, a longer research exploring the role of TAP and GOSIP in online behaviours could involve all the TAP subscales including the *Dependence* subscale. Although this study in no way addressed social media and internet addiction, it is worth adding this aspect into future research considerations. It is possible that helping others online may be one of the behaviors undertaken to satisfy the need to represent oneself on social media, or some way to rationalize the large number of hours spent online.

4.2 General Conclusions

This study occupies a gap in the literature by offering new insights into the role of human attitudes towards online interactions and technology in helping behaviours. Understanding what may limit some individuals' propensity to participate—in addition to which features might remedy their concerns—should assist in enhancing engagement in social initiatives. Online platforms carry unique potential in enabling ways to engage in altruistic behaviours for those who avoid social interactions. Facilitating methods of participation that meet varying individual needs may result in the inclusion of volunteers who are less socially engaged and are more anxious in their daily lives.

References

1. Website of the Republic of Poland. Wprowadzamy stan epidemii w Polsce (2020). https://www.gov.pl/web/koronawirus/wprowadzamy-stan-epidemii-w-polsce. Accessed 8 Sept 2020
2. Puto, K.: 7 doskonałych internetowych inicjatyw na czas pandemii (2020). https://krytykapolityczna.pl/felietony/kaja-puto/epidemia-koronawirus-organizujemy-sie/. Accessed 12 Sep 2020
3. Nowak, M.K.: Solidarność jak wirus. Rośnie w tempie wykładniczym. "Widzialna ręka" ma ponad 150 lokalnych grup (2020). https://oko.press/widzialna-reka-ma-juz-ponad-150-lokalnych-grup/. Accessed 12 Sept 2020
4. Jemielniak, D., Przegalinska, A.: Collaborative Society. MIT Press, Cambridge 18 February 2020
5. Bailey, A.A.: Factors promoting social CRM: A conceptual model of the impact of personality and social media characteristics. Int. J. Cust. Relat. Mark. Manage. (IJCRMM). 6(3), 48–69 (2015)
6. Liang, C.C., Dang, H.T.: Factors influencing office-workers' purchase intention though social media: an empirical study. Int. J. Cust. Relat. Mark. Manage. (IJCRMM). 6(1), 1–6 (2015)
7. Świdrak, J., Pochwatko, G., Matejuk, P.: Copresence and well-being in the time of Covid-19: is a video call enough to be and work together? In: Biele, C., Kacprzyk, J., Owsiński, J.W., Romanowski, A., Sikorski, M. (eds.) MIDI 2020. AISC, vol. 1376, pp. 169–178. Springer, Cham (2021). https://doi.org/10.1007/978-3-030-74728-2_16
8. Blazevic, V., Wiertz, C., Cotte, J., de Ruyter, K., Keeling, D.I.: GOSIP in cyberspace: conceptualization and scale development for general online social interaction propensity. J. Interact. Mark. 28(2), 87–100 (2014)
9. Magotra, I., Sharma, J., Sharma, S.K.: Assessing personal disposition of individuals towards technology adoption. Future Bus. J. 2(1), 81–101 (2016)
10. Burns, S., Roberts, L.: Applying the theory of planned behaviour to predicting online safety behaviour. Crime Prev. Community Saf. 15(1), 48–64 (2013). https://doi.org/10.1057/cpcs.2012.13
11. Ratchford, M., Barnhart, M.: Development and validation of the technology adoption propensity (TAP) index. J. Bus. Res. 65(8), 1209–1215 (2012)
12. Bosancianu, C.M., Powell, S., Bratović, E.: Social capital and pro-social behaviour online and offline. Int. J. Internet Sci. 8(1), 49–68 (2013)
13. Erreygers, S., Vandebosch, H., Vranjes, I., Baillien, E., De Witte, H.: Development of a measure of adolescents' online prosocial behaviour. J. Child. Media 12(4), 448–464 (2018)
14. Wright, M.F.: Predictors of anonymous cyber aggression: The role of adolescents' beliefs about anonymity, aggression, and the permanency of digital content. Cyberpsychology Behav. Soc. Network. 17(7), 431–438 (2014)

15. Wright, M.F., Li, Y.: The associations between young adults' face-to-face prosocial behaviours and their online prosocial behaviours. Comput. Hum. Behav. **27**(5), 1959–1962 (2011)
16. Oda, R., et al.: Personality and altruism in daily life. Pers. Individ. Differ. **1**(56), 206–209 (2014)
17. Ben-Ner, A., Kramer, A.: Personality and altruism in the dictator game: relationship to giving to kin, collaborators, competitors, and neutrals. Pers. Individ. Differ. **51**(3), 216–221 (2011)
18. Simpson, B., Willer, R.: Altruism and indirect reciprocity: The interaction of person and situation in prosocial behaviour. Soc. Psychol. Q. **71**(1), 37–52 (2008)
19. Lewis, K., Gray, K., Meierhenrich, J.: The structure of online activism. Sociol. Sci. **18**(1), 1–9 (2014)
20. Tsvetkova, M., Macy, M.W.: The social contagion of generosity. PLoS ONE **9**(2), e87275 (2014)
21. Lacetera, N., Macis, M., Mele, A.: Viral altruism? Charitable giving and social contagion in online networks. Sociol. Sci. **1**, 3 (2016)
22. McCroskey, J.C., Richmond, V.P.: Willingness to communicate: a cognitive view. J. Soc. Behav. Pers. **5**(2), 19 (1990)
23. Hochberg, Y., Benjamini, Y.: More powerful procedures for multiple significance testing. Stat. Med. **9**(7), 811–818 (1990)
24. Cattell, R.B.: Anxiety and motivation: theory and crucial experiments. Anxiety Behav. **1**, 23–62 (1966)
25. Spielberger, C.D., Rickman, R.L.: Assessment of state and trait anxiety. In: Anxiety: Psychobiological and Clinical Perspectives, pp. 69–83 (1990)
26. Negd, M., Mallan, K.M., Lipp, O.V.: The role of anxiety and perspective-taking strategy on affective empathic responses. Behav. Res. Ther. **49**(12), 852–857 (2011)
27. Todd, A.R., Forstmann, M., Burgmer, P., Brooks, A.W., Galinsky, A.D.: Anxious and egocentric: how specific emotions influence perspective taking. J. Exp. Psychol. Gen. **144**(2), 374 (2015)
28. Lam, S.Y., Chiang, J., Parasuraman, A.: The effects of the dimensions of technology readiness on technology acceptance: an empirical analysis. J. Interact. Mark. **22**(4), 19–39 (2008)
29. Nong, Z., Gainsbury, S.: Website design features: exploring how social cues present in the online environment may impact risk taking. Hum. Behav. Emerg. Technol. **2**(1), 39–49 (2020)

Deploying Enhanced Speech Feature Decreased Audio Complaints at SVT Play VOD Service

Annika Bidner[1] , Julia Lindberg[1] , Olof Lindman[1] ,
and Kinga Skorupska[2(✉)]

[1] SVT, Stockholm 105 10, Sweden
annika.bidner@svt.se

[2] Polish-Japanese Academy of Information Technology, XR Lab, Warsaw, Poland
https://www.svtplay.se/genre/tydligare-tal

Abstract. At Public Service Broadcaster SVT in Sweden, background music and sounds in programs have for many years been one of the most common complaints from the viewers. The most sensitive group are people with hearing disabilities, but many others also find background sounds annoying. To address this problem SVT has added Enhanced Speech, a feature with lower background noise, to a number of TV programs in VOD service SVT Play. As a result, when the number of programs with the Enhanced Speech feature increased, the level of audio complaints to customer service decreased. The Enhanced Speech feature got the rating 8.3/10 in a survey with 86 participants. The rating for possible future usage was 9.0/10. In this article we describe this feature's design and development process, as well as its technical specification, limitations and future development opportunities.

Keywords: Video streaming · User experience · Hearing loss · Participatory design · Sound design

1 Introduction and Related Works

SVT is the Public Service TV Broadcaster in Sweden. According to SVT's broadcasting licence [2] and audibility policy from 2020, SVT should prioritise good audibility by taking into account that background sound can make it harder for people with hearing loss to hear the program dialogue. The audibility work is fundamental for the public service offer to be accessible to everyone in Sweden. In 2020, about 1.5 million people in Sweden experienced hearing loss, 18% of the population over 16 [4]. A similar commitment is seen in the European Accessibility Act.

People with hearing disabilities are sensitive to background noise, but many others also find background sounds annoying, for example when loud sounds

---

Supported by The Swedish Association of Hard of Hearing People.

C. Biele et al. (Eds.): MIDI 2021, LNNS 440, pp. 277–285, 2022.
https://doi.org/10.1007/978-3-031-11432-8_28

could disturb neighbours, or when the viewing environment is noisy. Shirley et al. [8] discuss some of the problems with the experience of speech in broadcast audio and many of these points are also applicable to TV broadcasts and streaming. There are several other techniques to address the problem of audio accessibility. All programs at SVT get subtitles, except short clips and live broadcasts. But this won't help if the viewer has dyslexia or is visually impaired. Dubbing is used for foreign content in many countries, but this is mainly done in kids' shows in Sweden. A study from 2012 showed that most participants strongly dislike dubbing, and tend not to use it [6]. At SVT, background music and sounds in programs have for many years been the most common type of complaint from the viewers. These two quoted complaints represent common problems:

Complaint 1: *"Hello! I'm listening to the program on the Swedish military plane efforts in Afghanistan. I've written to you before about these constant background noises in your programs. This program is the worst so far, considering the annoying airplane noise and the "excuse me" goddamn piano tinkle. I don't know who handles my message but I know that someone responsible should listen to the program and try to realise how people with hearing loss experience this background sound. We have the right to perceive what is said in a program. I might contact The Swedish Association of Hard of Hearing People to get some improvements in this area."*

Complaint 2: *"WHY must the background music be so loud, AT THE SAME TIME as people are talking in a program???? My hearing is normal, but it's Impossible to hear what they are saying. Luckily, there are subtitles, so i can use that to understand what is said. The background noise is sooo annoying and distracting. It takes away the pleasure of viewing the show I want to see. Is the noise really necessary?"*

In 2019, SVT received more than a hundred sound complaints per week. SVT often adjust the sound mix of programs that get complaints, but this hasn't reduced the number of complaints. Even so, no real progress had been made in SVT Play, SVT's VOD service, to improve audibility of speech in TV programs, until the experimental project, described in this paper, started in January 2020.

The main hypothesis underlying the project was: If SVT can provide an audio feature that makes it easier to hear the dialogue, and include it in a substantial number of programs, many people will use it and the complaints to SVT on audibility will decrease.

An obvious alternative to enhancing the speech in videos would be to keep background sounds in programs to a minimum. But most viewers enjoy music and relevant sounds in programs. It helps create a more emotional and immersive experience [7]. Thus, BBC Research and Development work on enhanced audio tested on "Casualty" drama series [10] became an inspiration for the project. BBC tested an Accessible and Enhanced Audio Mix in the form of a slider on the Taster platform, in 2019 [3], but so far, it has not been released as a standard feature. Another source of inspiration was the German Fraunhofer institute, that produces technical solutions for enhancing the human voice and reducing background noise [5].

Fig. 1. Left to Right: Testing and iteration of the first prototype, January 10, 2020; Prototype 1: on/off; Prototype 2: Three steps; Prototype 3: Demo version with −15 and −30 dB

In the first phase of the project, SVT representatives contacted different stakeholders, and discussed current audibility research and possible technical directions to enhance speech in TV programs. First, ORCA Europe, a Hearing Research Laboratory in Stockholm. SVT considered strengthening the main speech frequencies to make the dialogue more prominent. Florian Wolters from ORCA Europe discouraged SVT from doing that: "Raising certain frequencies in the dialogue will make the sound unnatural. It should be as natural as possible." [9] Josefina Larsson, also from ORCA, said: "To lower the background sounds is a good idea. That should include all sounds that isn't notional dialogue." Alf Lindberg from SAHHP said "Avoid sudden changes in volume. Hearing aids adapt to a certain sound level, and users don't get a good experience when background noise is raised between dialogue passages." He promoted a continuous audio mix, to cater to different degrees of hearing loss. SVT has also been in contact with Dolby, a US company specializing in audio noise reduction and encoding. [1] Dolby's solution "Dialog enhancement" can be applied to stereo sound, but it could not separate speech in the way SVT was aiming for.

2   Methods

The first approach in the project was to manipulate the sound of video content by enhancing the main sound frequency bands (1–4 kHz) for human speech. A prototype was made and tested within the project group, but it was clear that this method did not fulfill its purpose. Background sounds within the voice frequency band were also enhanced and some important parts of speech with frequencies outside the span were lost. According to hearing experts at ORCA Europe, this solution was questionable, as it would distort the sound, so this approach was abandoned. The second approach was to use video with multi-channel audio (5.1 or 3.0 sound) and adjust the balance between the different channels in the downmixing process, with notional speech in the center channel. The recommendation from hearing experts was to enhance the speech relative to the background by decreasing the volume of background sounds. New prototypes were created, with either −3 or −10 dB background noise (Fig. 1).

Fig. 2. Left: UI used in SVT Play, September 2020; Right: Stereo waveform of regular downmix (Upper green) and Enhanced Speech downmix (Lower purple)

These prototypes were used in a participatory design session with representatives from the The Swedish Association of Hard of Hearing People along with developers, sound technicians, UX designers and accessibility specialists. It was clear that the reduction of background sound was not sufficient. The quality of the sound separation in channels as well as legal rights for manipulating sound in video content were issues that needed to be solved.

The term Enhanced Speech (Tydligare tal) was selected in cooperation with Tillgänglighetstruppen, an accessibility Facebook group with 2,000 members.

They also took part in user testing. The prototype UI was iterated after feedback from users. The two sound alternatives -15 dB -30 dB, were considered too complicated and technical, so those were merged into one. This had technical benefits, since only one extra audio track had to be produced for each episode. The result was this UI, used in production in svtplay.se (Fig. 2).

2.1   Technical Description

On a technical level, Enhanced Speech is achieved by utilizing an audio panning filter, to create a downmix matrix that converts predefined multichannel audio into a dialogue enhanced audio track. Specifically, the desired audio output is accomplished by modifying the digital gain of each individual audio channel in a remixing or downmixing process. The parameters used to decide the amount of gain is defined as decimal multiples rather than a dB values, and in effect the procedure alters the sound strength of each individual channel prior to the actual mixing. Before the modified input channels are merged together a programmatic check is made to measure if either of the resulting tracks are going to clip above 0 dBFS, and if that is the case the whole track will be normalized by max peak when mixed. The configuration of the panning filter is static, which means that a single downmix matrix is used for all content and that the parameters are always the same. For multichannel to stereo conversions the matrix is as follows (Fig. 3):

$$L_{ES} <= (0.25 * L) + (1.5 * C) + (0.25 * Ls)$$

$$R_{ES} <= (0.25 * R) + (1.5 * C) + (0.25 * Rs)$$

whereas a normal downmix from 5.1 surround to 2.0 stereo, as defined in the EBU R 128 recommendation (https://tech.ebu.ch/docs/r/r128.pdf), would use the following values:

$$L_{Stereo} = (1 * L) + (0.707 * C) + (0.707 * Ls)$$

$$R_{Stereo} = (1 * R) + (0.707 * C) + (0.707 * Rs)$$

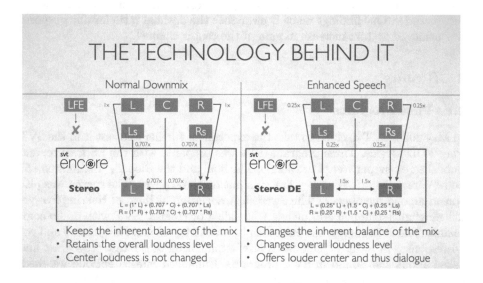

Fig. 3. Downmix matrix

When creating a 5.1 Enhanced Speech track the matrix is:

$$L_{ES} = 0.4 * L$$

$$R_{ES} = 0.4 * R$$

$$C_{ES} = 1.5 * C$$

$$LFE_{ES} = 0.25 * LFE$$

$$Ls_{ES} = 0.25 * Ls$$

$$Rs_{ES} = 0.25 * Rs$$

Furthermore, the filter relies entirely on the assumption that the input audio streams have a standard channel layout, which is to say:
3 channels: L, R, C
6 channels: L, R, C, LFE, Ls, Rs
8 channels: L, R, C, LFE, Ls, Rs, StereoLeft, StereoRight
   Although the parameter values for the panning filter might seem intuitive, they are far from arbitrary, since they were devised through empirical testing.

Initially we assumed that it would be sufficient to reduce the gain of every channel except the center, the idea being that by lowering the volume of everything else, the dialogue would stand out psychoacoustically. However, the effect turned out to be insufficient as the dialogue was still perceived as hard to hear. So, we created several variants of the same set of media clip, with different parameter values respectively, and made them available among our public test videos (https://www.svtplay.se/testbild). From there, we could easily test the results on several devices and in various viewing settings using a wide enough range of test material. Our findings made it clear that the optimal ratio for our purposes was achieved at 1.5x increase in gain of the center channel.

3   Results

3.1   First Programs

In May 2020, SVT decided to take the experiment further and test it in the SVT Play VOD service. Three program titles was prepared with the ES feature, and published on svtplay.se: "Caravaggio, The Soul and the Blood", "Sommaren 85" and "Värsta listan". A survey link was put on these pages. The survey was also spread among members of The Swedish Association of Hard of Hearing People, Tillgänglighetstruppen and people who called or emailed SVT with audio complaints. The 2020 survey was published between May 13 and November 9 and got 86 responses in total. It included a brief explanation of the test procedure, and asked about age, sound in SVT programs, Enhanced Speech episode watched, opinion on and differences regarding the default audio and the ES audio in the episode, opinion about the ES feature and the likelihood of using the feature in the future.

The result was very positive, in all age groups. Enhanced Speech got the rating 8.3/10, compared with the default sound, which got 5,6/10. People said the difference between the sound was 8.2/10 and on the question whether they would use ES if it was available on every SVT program, they answered 9,0/10. As a comparison, Spoken Subtitles (Uppläst undertext in Swedish), was released in SVT Play in April 2021. A similar survey was done, with the main question: What is your opinion of the Spoken subtitles? The result was 5.8/10, with 21 participants. So, SVT's visitors gave ES a much higher rating compared to another new and anticipated accessibility feature. The survey participants commented for example: "Wonderful! Finally, I can watch TV without having to use subtitles.", "I don't need as high volume to hear what is said, so I don't disturb my neighbours.", "Excellent for me, I have severe tinnitus." or "An option to use when you have a bad speaker."

3.2   Scaling up

Because of the positive feedback from the audience, SVT continued to produce more programs with ES. The standard contracts for buying or ordering new shows was updated to include requirements on multi-channel sound. The internal sound mixing process was changed, to make sure that the center channel included only the voice. The coding process of the multi-audio programs was automated from the beginning of May 2021, so that SVT could offer many more titles each week. All shows with 5.1 or 3.0 sound now gets Enhanced Speech, if SVT doesn't manually exclude them. On June 21 2021, there were 128 shows with the ES feature available, out of around 2,300 in total.

Follow-up Survey. A follow-up survey was conducted between May 4 and June 21 2021, after the automated production of ES programs had started. This survey had 20 participants, and the result was similar to the first one, with a score of 8.4/10 score for the ES feature, and 8.7/10 on the probability of future usage.

Effect on Audio Complaints. As the number of titles with alternative audio has increased, the video starts with Enhanced speech have gone up by 431%, from 2,721 to 14,452 per week, and the complaints have gone down by 91%, from 104 to 9 per week, from April 12 to June 27 (Fig. 4).

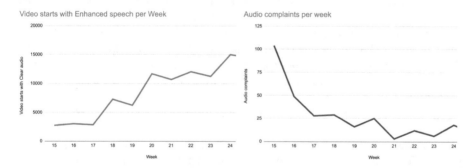

Fig. 4. Left: Video starts with Enhanced speech per week, May-June 2021. Right: Audio complaints per week, May-June 2021

3.3   Limitations

The Enhanced Speech process assumes that all of the notional dialogue resides in the center channel, which wasn't always the case. The blurred center channel greatly reduced the benefit of the ES feature. But from May 2020 onwards, SVT reserved the center channel for notional speech in new programs. Another limitation is the static nature of the mix and lack of dynamic processing. Since all gain levels are adjusted as per the downmix matrix, the sections of a program that completely lack dialogue become very quiet. This effect gets stronger in some programs, depending on the balance of speech and background sound. By design, the Enhanced Speech mix warps the creative intent of an audio mix to a lesser or greater extent, and although the actual effect can still be a net benefit to the viewer, it is nonetheless important to emphasize this distortion. Moreover, the conversion process can only use multichannel audio as input, the procedure does not handle mono to mono or stereo to stereo. This means only multichannel audio programs got coded with Enhanced Speech. During 2020 and 2021, these programs were high-end and popular productions. This might have skewed the experience of these shows in a positive way, compared to others, with stereo sound.

ES is so far only available for VOD programs in SVT Play, not on broadcast, live streams or online channels, due to technical limitations. This means that some programs that get many audio complaints, like News and Sports, were not included. Also, SVT decided to avoid ES in programs based on music and singing, because it can give an unpleasant experience to hear the voice without musical instruments. Regarding the target group, the surveys were open to anyone who visited play.se, and there wasn't any background question regarding hearing, so it's impossible to tell if the participants were representative of the main target group, people with hearing loss.

4 Conclusions and Future Work

More research on Enhanced Speech in TV programs is needed. So far, SVT's main target group has been the hard of hearing community, and the Enhanced Speech feature was tailored to be compatible with hearing aids, and used multi-channel sound. This direction could be explored further, by adapting the mixing levels for different program genres, refining what sounds should be present in the center channel and letting users try out multiple levels and variants of static downmixing.

For people not using hearing aids, two alternative solution types could be explored, Dynamic Downmixing and Parametric Equalization (EQ). Dynamic downmixing means background noise is lowered when a recognizable voice is present. This technique is already used in the Spoken Subtitles feature at SVT. With Parametric EQ, frequencies present in the voice range are boosted. Both of these techniques involve the problem of recognizing what actually is a voice, and AI could be a big help here.

Even if sound complaints have decreased radically at SVT, some remain. They often concern News and Sports, and other programs with stereo sound. That means solutions for stereo sound, as those mentioned above, are important to consider for SVT. These solutions might not reach the same speech intelligibility as in SVT's version of Enhanced Speech for multi-channel sound, but they could still bring a lot of value for viewers of the stereo sound content. Another area of exploration is how sound alternatives should be presented in the UI to be discovered by the users who need them. SVT is open for collaborations in all areas mentioned above, that may improve the users' viewing experience.

The solutions explored in this paper could also be applied to adjacent industries, for example radio, podcasts and computer games. It is our hope that they will contribute to the creation of a more accessible and enjoyable soundscape in people's lives.

References

1. Predicting impact of dialog enhancement on large audiences. https://dolby.io/blog/dialog-enhancement-on-large-audiences
2. Broadcasting license for Swedish television 2020–2025 (2019). https://www.mprt.se/globalassets/dokument/sandningstillstand/public-service/sandningstillstand-svt-2020-2025.pdf
3. Casualty, loud and clear - our accessible and enhanced audio trial - BBC R&D, August 2019. https://www.bbc.co.uk/rd/blog/2019-08-casualty-tv-drama-audio-mix-speech-hearing
4. The Swedish association of hard of hearing people: hearing loss in numbers (2020). https://hrf.se/app/uploads/sites/13/2020/12/Statisik-2020-HRF-och-SCB-Horselbron-och-forskning.pdf
5. Speech intelligibility for broadcasting - Fraunhofer IDMT, July 2021. https://www.idmt.fraunhofer.de/en/institute/projects-products/projects/SI4B.html
6. Fors, S.: The attitudes of Swedish viewers regarding the quality of translation in audiovisual media. Ph.D. thesis (2012)
7. Hoffman, R.: What is the function of film music? January 2020. https://www.robin-hoffmann.com/tutorials/what-is-the-function-of-film-music
8. Shirley, B., Ward, L.: Intelligibility versus comprehension: understanding quality of accessible next-generation audio broadcast. Univers. Access Inf. Soc. **20**, 691–699 (2021)
9. Smeds, K., Wolters, F., Rung, M.: Estimation of signal-to-noise ratios in realistic sound scenarios. J. Am. Acad. Audiol. **26**(2), 183–196 (2015)
10. Ward, L., Paradis, M., Shirley, B., Russon, L., Moore, R., Davies, R.: Casualty accessible and enhanced (A&E) audio: trialling object-based accessible TV audio. AES Convention: 147 (2019)

Brief Overview Upper Limb Rehabilitation Robots/Devices

Mariusz Sobiech[1,2,3](✉) ⓘ, Wojciech Wolański[2] ⓘ, and Ilona Karpiel[1] ⓘ

[1] Łukasiewicz Research Network—Institute of Medical Technology and Equipment, 118 Roosevelta Street, 41-800 Zabrze, Poland
mariusz.sobiech@itam.lukasiewicz.gov.pl

[2] Department of Biomechatronics, Silesian University of Technology, 40 Roosevelta Street, 41-800 Zabrze, Poland

[3] PhD School, Silesian University of Technology, 2a Akademicka Street, 44-100 Gliwice, Poland

Abstract. The rehabilitation approach has changed with the appearance of robots. As a results the rehabilitation costs significantly decrease but also time for both the patient [1], who does not have to commute for long time to the office and medical professionals. Nowadays medicine, computer science, electronics, and engineering, in general, are strongly connected. A group of specialists is working on newer and newer solutions to improve both diagnosis and therapy. This article provides an overview of basic rehabilitation robotic solutions used in the rehabilitation of upper limb functions.

The literature used is based on PubMed and Scopus databases included articles published between 1999 and 2021. Eligibility criteria included upper limb exoskeletons for rehabilitation of both the wrist, elbow, and shoulder joints.

This paper provides an overview of an important research subject and highlights the current knowledge in the field. Despite extensive attempts to develop rehabilitation systems, exoskeletons are primarily uncommercialised despite a large number of prototypes.

Keywords: Upper limb exoskeleton · Robots rehabilitation · Limb functions upper · Drive units · Kinematic chain of exoskeleton · Review of robots

1 Introduction

We live in an aging society, with an increasing number of diseases, in particular CNS diseases, and with other diseases leading to impaired efficiency, such as atherosclerosis, diabetes, osteoarthritis, etc. Taking into account the CNS, at least 450 million people worldwide suffer from neurodegenerative diseases (around 50 million people suffer from neurodegenerative diseases). Other brain diseases include i.e. stroke (15 million), traumatic brain injury (TBI), and brain tumors, which affect about 1.5 million people. In total, it is about 0.5 billion people affected by brain diseases. According to WHO, the third on the list of civilization diseases leading to disability is stroke. An indispensable element in the process of treating strokes is rehabilitation, the effectiveness of which is very strongly dependent on the time of its implementation after the occurrence of a stroke. Treatment of these diseases and conditions, which largely contribute to motor impairment, is lengthy (situation-dependent), costly and involves many people, including healthcare professionals and family members.

Currently, drug treatment in the form of injections, suppositories, tablets and ointments is the most common. This treatment often leads only to a reduction of pain and only partial recovery. Moreover, the effects of such treatment are not permanent. It is impossible to imagine modern medical treatment without rehabilitation and physiotherapy, whose aim is to restore full and permanent functional capacity. It must be admitted that sometimes it is necessary to carry out surgical - orthopedic treatment, which, however, in order to achieve full success requires the effect of postoperative rehabilitation.

Performing therapeutic exercises requires great commitment from the physiotherapist and is very time consuming. To achieve the expected effect the exercises have to be repeated many times individually (patient - therapist). Group exercises are more beneficial in organisational and economic terms, but unfortunately they are not equivalent and do not lead to the expected effects of therapy. The constant repetition of the therapeutic movement sequences leads to the therapist's weariness, which may result in less accurate execution of the exercise or shortening the duration of the exercises. The solution to the problem may be the use of robots to relieve the physiotherapist from monotonous and exhausting physical work, at the same time allowing for the implementation of trainings with many patients by one physiotherapist. In addition, a rehabilitator using a robot obtains a diagnostic tool, because the robot's sensors can, for example, measure ranges of mobility in a given joint, or the strength of selected muscles.

In this review the literature used is based on PubMed and Scopus databases including articles published between 1999 and 2021. A search was used based on the following keywords: "upper limb", "robot" "rehabilitation". The total number of results was 1700, including 156 reviews. The review was narrowed down to full text of publications available without charge and review papers and systematic reviews, which numbered 94. Next, the database was searched using the keywords "(exoskeleton) AND (upper limb)", and the area was also narrowed down to full text of publications, the number of which was 20. From among the available articles, only those focused on the presentation of exoskeletons that can be rehabilitated in all 3 joints (wrist and hand joints, elbow joint, shoulder and clavicle joint) were selected.

As it turns out, the information found in the searched database would not allow the presentation that would be 100% satisfactory to the readers, because key technical information is often not described in publications. Taking into account, for example, one of the main parameters that distinguishes selected robots, i.e. degrees of freedom, it was necessary to additionally search manufacturers' websites or additional materials found on the Internet, which made it difficult to create the review.

1.1 The Importance of Anthropometric Values for the Upper Limb

The upper limb plays a very important role in human daily life. It enables people to perform grabbing and cognitive activities. Due to the number and variety of tasks performed, the upper limb is particularly vulnerable to injury. This is the reason of study for physicians and physiotherapists as well as biomechanists [2–4].

One of the main research conducted on the upper limb are anthropometric tests, whose purpose is to provide an objective and accurate data, which are used to create rehabilitation equipment. The mentioned measurements: total limb length - measured from the acromion process to the styloid process of the elbow bone or the end of the middle finger; arm length - measured from the acromion process to the lateral epicondyle of the humerus; forearm length - measured from the ulnar process to the epicondyle process of the ulnar bone; and circumferences of the upper limb segments. As we know, the upper limb is characterised by a large range of motion, which results directly from its anatomical structure. In the shoulder joint, three movements take place on three planes: the sagittal plane is flexion - straightening (in the range of 60° of expansion and 180° of flexion), the coronal plane is abduction - adduction (in the range of 90° of abduction, 180° of abduction with the shoulder blade, 20° of adduction) and the transverse plane is external rotation - internal rotation (in the range of 98° of internal rotation and 90° of external rotation). In the other joints of the upper limb, movement takes place in only two planes. In the elbow joint, the movement is in the sagittal plane, in which flexion-straightening (150° flexion and 0° extension), and in the transverse plane, which is a rotation of the forearm (80° supination and 90° pronation). At the wrist, the movement takes place in the sagittal plane, flexion-extension (palmar flexion 70° and dorsiflexion 80°), and in the frontal plane, inversion-adduction movement (radial flexion 20° and ulnar flexion 40°).

2 Robots

The world is changing, therefore there is a need for new inventions, robots, and new solutions to support treatment. An aging population presents a new challenge. A challenge when it comes to treatment.

Scientists have found that particular activities of the brain can be transferred to a different location in the brain, and this is known as neuroplasticity. Repetitive motions for the impaired limbs allow the brain to develop new neural pathways and, ultimately, restore full or partial control of motor functions. Using a rehabilitation robot could trigger neuroplasticity by providing a repetitive exercise for the impaired functions.

The authors decided to present selected robots and briefly describe the construction and operation.

2.1 ARMin

In 2007 the first multi-armed upper limb rehabilitation device ARMin was developed [5, 6]. It was developed at the University of Zurich in cooperation with the Hocoma company and therapeutic doctors from Zurich's Balgrist clinic. Originally, it featured six degrees of freedom, four of which were propelled and the other two were passive. In that way, a kinematic scheme from the shoulder joint to the forearm was realised. The next version added two more degrees of freedom to allow movement of the forearm and wrist. The device is driven by Maxon RE series motors, which are DC motors with graphite brushes. The motors are paired with harmonic gears. An interesting feature of this device is the solution providing internal and external rotation of the shoulder. It is achieved by a special rotating module, made of two semi-cylinders. The inner element is guided by ball bearings mounted in the outer element. The drive is realised by steel cables mounted to the ends of the inner half-cylinder, which roll over the motor drive shaft. It should also be mentioned that the device can adapt to different lengths and sizes of the upper limb. Although the device in this form allows performing almost all basic and complex exercises, it does not allow to perform Proprioceptive Neuromuscular Facilitation (PNF) exercises due to the limited range of its performance.

2.2 ArmeoPower

Based on ARMin, the ArmeoPower (https://www.hocoma.com) device was developed in 2011, which is one of the first commercialised robots designed for upper limb rehabilitation. The device is intended for patients who have completely lost or have significant reduction of functionality of the upper limb due to neurological problems or injuries of the nervous system. The device has six degrees of freedom, where each degree of freedom is equipped with an independent motor and two force sensors. The device can be adapted to the patient thanks to the adjustment of the column height and length of the arm and forearm parts of the exoskeleton. The device is capable of performing the movements as shown in Table 1.

Table 1. Technical data ArmeoPower (version 1.0).

Horizontal shoulder abduction	$-169°$ to $+50°$
Shoulder flexion/extension	$+40°$ to $+120°$
Shoulder internal/external rotation	$0°$ to $90°$
Elbow flexion/extension	$0°$ to $100°$
Forearm pro-/supination	$-60°$ to $60°$
Wrist flexion/extension	$-60°$ to $60°$

The robot continues to be refined and successfully used in clinical trials. A recent paper by Meyer et al. [7] presents to assess feasibility, safety, and potential efficacy of a new intensive focused arm-hand BOOST program and to investigate whether there is a difference between early vs. late delivery of the program in the sub-acute phase post stroke.

2.3 Armeo®Spring

Armeo®Spring is another device for upper-limb neurorehabilitation from the Armeo® family. Armeo is a commercial replica of the T-WREX device that was developed in the USA in 2004 [10]. The rehabilitation is based on working with an orthosis (exoskeleton) whose system of springs supports the rehabilitated limb and supports training. The orthosis is designed for patients with limited or lost arm function. Dysfunctions caused by injuries to the central or peripheral nervous system are treated by training that includes exercises to increase muscle strength, range of motion of the limb, and motor skills. This device has five degrees of freedom, where three are at the shoulder joint, and one each at the elbow and wrist joints. The device is characterized by the fact that it is a passive device with no drives, but it has an advanced spring mechanism that relieves the upper limb during exercises and supports training. There is also Armeo Spring Pediatric [8] - a version designed for children who require rehabilitation of the upper limb. It is based on the Armeo Spring design but the length of the brace and strain relief are adapted to the needs of children aged 4 to 12 years old. Recently, new techniques based on robotic-assistive devices have been increasingly beneficial [9]. The latest research has shown that even a short-term, two-week training program with new technologies had a positive effect and significantly recovered Stroke Patients functional level in self-care, upper limb motor ability (dexterity and movements, kinematic data, grip strength), visual constructive abilities (memory, visuo spatial abilities, attention, and complex commands).

Armeo® devices increase the effectiveness and intensity of the therapy by including even chronically ill patients self-initiated movements and motivate them to train with high intensity during the rehabilitation process [11].

2.4   Renus

The next device is the Polish project called Renus-1. It is a mechatronic system supporting motoric rehabilitation of the upper limb in patients after strokes or orthopedic diseases. The device was realised as a project in Industrial Research Institute for Automation and Measurements PIAP in Warsaw, coordinated by Institute of Exploitation Technology from Radom in the years 2006–2010 and 2013–2014 [15]. The system consists of a mechanical part - manipulator, control system, and software. The manipulator makes it possible to create a spatial trajectory of motion of the patient's hand and upper limb. The manipulator arm consists of two rigid elements connected by joints and is articulated to the ambulance moving on a vertical sled attached to a fixed column. The articulation axes of the manipulator's arm are vertical. From the kinematics point of view, the mechanical structure of the device is a mechanism with three degrees of freedom, which allows the hand grip to be moved up/down, left/right, to/from each other. At the end of the arm there is the mechanical interface of the manipulator equipped with a multi-axis force and torque sensor. The drive system is based on three servo drives, which are synchronous motors from Mitsubishi Electric. The motors have integrated 17-bit encoders and cooperate with planetary gears of Alpha company. The largest of them is responsible for the Z-axis drive and has a power of 100W while the other two have a power of 50W. RENUS is described in detail in 3 items [12–14], where you will find a detailed description of the system design or software. The RENUS system has been tested for its therapeutic purposes and performance properties under domestic conditions.

There is also a version of the device for rehabilitation of the lower limb known as Renus-2.

2.5 ALEx

A representative of an advanced rehabilitation robot that enables the implementation of training multifaceted is also a device called ALEx by Kinetek, developed in 2013 at PERCRO Lab in Italy. The robot enables operation in a configuration for one or two arms simultaneously. Each arm is equipped with four active degrees of freedom equipped with actuators and two passive ones equipped only with sensors. Four BLDC brushless motors are with integrated optical incremental encoders. In addition, the device is equipped with absolute angle encoders, which are mounted directly at the point of rotation. The unique feature of this design is the patented implementation of an arm rotation mechanism that uses a remote rotation center. The movement from the motor to the driven connection is realized using a linkage gear. The arm of the exoskeleton weighs only 4.5 kg. ALEx device is a medical device with CE class IIa certification and can operate in 3 modes (passive, assistive and assisted when needed). In passive mode the patient moves the upper limb, and the robot measures the movements. In assistive mode the robot guides the patient's upper limb. In the so-called "assisted when needed" mode, the robot guides the rehabilitated person's arm to the target position if the user does not initiate the movement in less than three seconds.

2.6 Harmony

An interesting project of an exoskeleton used for rehabilitation of upper limbs mainly after stroke is the Harmony device. The work on it began in 2011 at the University of Texas, USA. It is the first-ever rehabilitation robot capable of rehabilitating both arms simultaneously. Each arm has seven active degrees of freedom, and a total of 14. The robot is equipped with SEA (series elastic actuators) drives, based with brushless DC motors Maxon Motor (EC flat series) combined with Harmonic Drive wave gears. Additionally, the device is equipped with four multi-axis force and torque sensors. Exoskeleton segment lengths can be customized for the individual patient.

2.7 IntelliArm

The IntelliArm is an exoskeleton designed and developed in 2007 in the USA. It is designed for upper limb rehabilitation of patients with neurological disorders. The project is based on MIT-Manus device developed at Boston Institute of Technology in 1997. The device has seven active degrees of freedom: four at the shoulder joint responsible for abduction/adduction, flexion/extension, internal/external rotation, and vertical movement of the shoulder joint. The next two degrees of freedom are at the elbow joint and one degree at the wrist [16]. In addition, the device has two passive degrees of freedom, which allow for posterior/anterior as well as medial and lateral displacement of the shoulder joint. Three multi-axial force sensors are mounted on the exoskeleton at each joint. Interesting mechanisms were used while designing of the device. Adduction/abduction and bending/straightening movement of the shoulder joint is transmitted from the actuator via cables. In the case of bending/straightening of the shoulder joint, the motor shaft is connected to a drum by a set of two cables, and another set of cables transmits the movement from the drum to the bending axis. The cables are tensioned by a tensioner. The arm and forearm rotation is performed by a mechanism using circular guides and a cable mechanism. Another interesting solution in this device is the wrist drive mechanism. The axis of the motor has been tilted 90 degrees using a bevel gear mechanism. Two cables wrap in the opposite direction around the motor axis are respectively mounted to drums 1 and 2. In the case of this device, it was possible to find information not only about the ranges of motion but also information about the maximum speeds and torques occurring during performing specific motions.

2.8 Aramis

Another device for upper limb rehabilitation is the Aramis system, developed in 2007 in Italy. It consists of two symmetric exoskeletons that interact with each other [17].

Each exoskeleton has six degrees of freedom and adjustment mechanisms allowing the exoskeleton to be perfectly adjusted to the arm length, forearm length, and height of the patient. The device allows for operation in a mode where the healthy limb forces the movement of the exoskeleton, which is replicated on the second exoskeleton, which forces the movement of the limb with paresis. The robot is built using DC brush motors with planetary gears Maxon Motor and optical encoders.

2.9 BONES

BONES developed at the Biomechatronics Lab - University of California Irvine. is an upper limb rehabilitation device with four degrees of freedom pneumatically actuated [18]. BONES is based on a parallel mechanism that moves the upper arm by means of two passive sliding rods that rotate relative to a fixed structural frame. Four mechanically grounded pneumatic actuators are located behind the main structural frame to control the movement of the arm through the sliding rods, while a fifth cylinder located on the structure was used to control elbow flexion/extension. The device supports a wide range of human arm motion while achieving low inertia and the ability to generate force directly on the arm. A key achievement of this is the ability to generate internal/external rotation of the arm without any circular bearing element.

2.10 ARM-100, ARM-200

Parallel to the above mentioned robots, in the years 2007–2009 in the Institute of Medical Technology and Equipment in Zabrze, a project of a device for multi-surface rehabilitation of the upper limb ARM-100 was realized (Fig. 1). The device was created in cooperation with the Upper Silesian Rehabilitation Centre "Repty" and Silesian University of Technology. The ARM-100 robot was created to support the rehabilitation of people with paresis of the upper limbs after diseases such as stroke, central nervous system injuries or rheumatoid arthritis. Rehabilitation with the ARM-100 robot is based on the PNF method (proprioceptive neuromuscular movement training). The device has seven degrees of freedom and allows for rehabilitation of the whole limb, both in the shoulder joint and in the elbow and wrist joints. A training session using the ARM-100 robot consists of two stages. The first stage involved "teaching" the robot the rehabilitation movement. In this stage, after gripping the patient's hand in the device, the rehabilitation therapist performs the rehabilitation movement by guiding the patient's hand. Based on signals from force sensors measuring the pressure exerted by the patient's limb on elements of the robotic arm, the device moves appropriate elements of its structure and, at the same time, memorizes their successive positions in the computer system. In the next stage of training, the robot reproduces the memorised model movement with the required speed and number of repetitions, testing at the same time whether the acceptable forces on the patient are not exceeded.

In addition to the passive rehabilitation described above, the device has also been designed for active rehabilitation with resistance, in which the patient himself carries out the movements and the device can put up defined resistance.

Work is currently underway to design a new ARM-200 device. The device is expected to have a greater range of motion and be ready for commercialisation.

Fig. 1. ARM-100. Own source (www.itam.lukasiewicz.gov.pl)

A comprehensive review on robot-assisted therapy for handtreatment can be found in Lum et al. [19] whether e.g. Babaiasl [20] and Kim et al. [21]. In other publications [22], the authors Heo et al. presented a broad survey on hand exoskeleton innovations for rehabilitation and assistance. Unfortunately, most of these devices have low wearability. Piazza et al. [23] have investigated novel solutions for assistive robotic tools to be used at home by chronic stroke patients. However, despite the large number of emerging solutions, only a small part is used in rehabilitation. It seems that the topic of how to speed up the deployment process and how to enable the safety usage of more and more robots would need to be addressed.

3   Summary

Publications indexed in the PubMed database on upper limb function improvement were reviewed. Table 2 summarizes the characteristics of the rehabilitation robots.

Despite a large number of publications on the rehabilitation of upper limb function, the number of papers on the usage of rehabilitation robots for this purpose is very small.

Table 2. Summarizes the characteristics of the rehabilitation robots.

Robot's name	Institute	Year	DoF(s)	Actuators	Application/Commercialized
ALEx	Kinetek PERCRO Lab, San Giuliano Terme, Italy	2013	4 active 2 passive	DC brushless motors, cable transmission, patented shoulder rotation mechanism	Post-stroke rehabilitation of one or two upper limbs/yes
Aramis	Istituto S.Anna, Italy	2009	6/12 active	DC brush motors with planetary gears Maxon Motor	Symmetrical rehabilitation of two upper limbs
ARM-100	ITAM Zabrze, Poland	2009	7 active	DC motors with planetary gears Maxon Motor	Multi-surface rehabilitation of the upper limb using the PNF method
ARMin (I)	ETH Zürich (Institut für Automatik), Switzerland	2007 (2001)	4 active 2 passive	RE series DC motors (Maxon Motors) with harmonic gears	Upper limb rehabilitation: arm and elbow (without PNF method)
Armeo Power	Hocoma, Switzerland	2011	6 active	DC motors with two angular sensors each	Rehabilitation for patients who have lost the function or have restricted function in their upper extremities/yes
Armeo Spring (be based T-WREX)/Armeo Spring Pediatric (for children)	Hocoma, Switzerland	2004 (T-WREX)	5 passive	System of springs (no drives)	Upper limb neurorehabilitation for adults and children/yes
Bones	BioRobotics Lab, University of California, USA	2013	4+2 active	Pneumatic actuators	Post-stroke rehabilitation of upper limb
Harmony	University of Texas, USA	2011	7/14 active	EC flat series DC motors (Maxon Motors) with harmonic gears (Harmonic Drive)	Rehabilitation of both upper limbs at the same time

(*continued*)

Table 2. (*continued*)

Robot's name	Institute	Year	DoF(s)	Actuators	Application/ Commercialized
IntelliArm (be based MIT-Manus)	USA (Boston Institute of Technology, USA)	2007 (1997)	7 active 2 passive	DC motors, cable mechanism	Upper limb rehabilitation of patients with neurological disorders
Renus-1	PIAP Warsaw and ITeE Radom, Poland	2006–2010/ 2013–2014	3 active	Servo drives: synchronous motors (Mitsubishi Electric) with planetary gears (Alpha)	Upper limb rehabilitation in patients after stroke or orthopedic diseases

Considering that rehabilitation robotics has been developing rapidly over the last 20 years and may represent a breakthrough in upper limb function rehabilitation, publications in this area still represent a small percentage. As well as the number of robots used for clinical research and rehabilitation is appallingly small.

Acknowledgements. This research was financed by the National Centre for Research and Development, Poland, under grant No POIR.01.02.00-00-0014/17 and was co-financed by the Ministry of Education and Science of Poland under grant No DWD/3/7/2019 - RJO15/SDW/001.

Conflict of Interest. The authors have no conflict of interest to declare.

References

1. Huang, V.S., Krakauer, J.W.: Robotic neurorehabilitation: a computational motor learning perspective. J. Neuroeng. Rehabil. **6**, 5 (2009). https://doi.org/10.1186/1743-0003-6-5
2. Guzik, A., Michnik, R., Rycerski, W.: The estimation of rehabilitation Progress in patients with psychomotor diseases of upper limb based on modeling and experi-mental research. Acta Bioeng. Biomech. **8**(2), 79–87 (2006)
3. Van Andel, C.J., Wolterbeek, N., Doorenbosch, C.A.M., Veeger, D., Harlaar, J.: Complete 3D kinematics of upper extremity functional tasks. Gait Posture **27**, 120 (2008)
4. Tejszerska, D., Świtoński, E., Gzik, M.: Biomechanika narządu ruchu człowieka. Wydawnictwo Naukowe Instytutu Technologii Eksploatacji -PIB, Radom (2011)
5. Nef, T., Mihelj, M., Riener, R.: ARMin: a robot for patient-cooperative arm therapy. Med. Bio. Eng. Comput. **45**, 887–900 (2007). https://doi.org/10.1007/s11517-007-0226-6
6. Nef, T., Colombo, G., Riener, R.: ARMin – Roboter für die Bewegungstherapie der oberen Extremitäten (ARMin – Robot for Movement Therapy of the Upper Extremities). auto. **53**, 597–606 (2005). https://doi.org/10.1524/auto.2005.53.12.597
7. Meyer, S., Verheyden, G., Kempeneers, K., Michielsen, M.: Arm-hand boost therapy during inpatient stroke rehabilitation: a pilot randomized controlled trial. Front. Neurol. **12**, 652042 (2021). https://doi.org/10.3389/fneur.2021.652042

8. Roberts, H., et al.: Constraint induced movement therapy camp for children with hemiplegic cerebral palsy augmented by use of an exoskeleton to play games in virtual reality. Phys. Occup. Ther. Pediatr. **41**, 150–165 (2021). https://doi.org/10.1080/01942638.2020.1812790
9. Adomavičienė, A., Daunoravičienė, K., Kubilius, R., Varžaitytė, L., Raistenskis, J.: Influence of new technologies on post-stroke rehabilitation: a comparison of armeo spring to the kinect system. Medicina **55**, 98 (2019). https://doi.org/10.3390/medicina55040098
10. Housman, S.J., Le, V., Rahman, T., Sanchez, R.J., Reinkensmeyer, D.J.: Arm-training with T-WREX after chronic stroke: preliminary results of a randomized controlled trial. In: 2007 IEEE 10th International Conference on Rehabilitation Robotics. pp. 562–568. IEEE, Noordwijk, Netherlands (2007). https://doi.org/10.1109/ICORR.2007.4428481
11. Kleim, J.A., Jones, T.A.: Principles of experience-dependent neural plasticity: implications for rehabilitation after brain damage. J. Speech Lang. Hear. Res. **51** (2008). https://doi.org/10.1044/1092-4388(2008/018)
12. Dunaj, J., Klimasara, W.: Rozwiązania sprzętowe i programowe w sterowaniu robotami rehabilitacyjnymi Renus. Pomiary Automatyka Robotyka, pp. 100–115 (2014)
13. Klimasara, W., Dunaj, J., Stempnik, P., Pilat, Z.: Zrobotyzowane systemy RENUS-1 oraz RENUS-2 do wspomagania rehabilitacji ruchowej po udarach mózgu. Pomiary Automatyka Robotyka. **2**, 577–589 (2009)
14. Mikołajewska, E., Mikołajewski, D.: Wykorzystanie robotów rehabilitacyjnych do usprawniania. Niepełnosprawność – zagadnienia, problemy, rozwiązania (2013)
15. Dunaj, J., Klimasara, W.J., Pilat, Z.: Human-robot interaction in the rehabilitation robot renus-1. In: Szewczyk, R., Kaliczyńska, M. (eds.) SCIT 2016. AISC, vol. 543, pp. 358–367. Springer, Cham (2017). https://doi.org/10.1007/978-3-319-48923-0_39
16. Park, H.-S., Ren, Y., Zhang, L.-Q.: IntelliArm: an exoskeleton for diagnosis and treatment of patients with neurological impairments. In: 2008 2nd IEEE RAS & EMBS International Conference on Biomedical Robotics and Biomechatronics. pp. 109–114. IEEE, Scottsdale, AZ, USA (2008). https://doi.org/10.1109/BIOROB.2008.4762876
17. Cerasa, A., et al.: Exoskeleton-robot assisted therapy in stroke patients: a lesion mapping study. Front. Neuroinform. **12**, 44 (2018). https://doi.org/10.3389/fninf.2018.00044
18. Milot, M.-H., et al.: A crossover pilot study evaluating the functional outcomes of two different types of robotic movement training in chronic stroke survivors using the arm exoskeleton BONES. J. NeuroEng. Rehabil. **10**, 112 (2013). https://doi.org/10.1186/1743-0003-10-112
19. Lum, P.S., Godfrey, S.B., Brokaw, E.B., Holley, R.J., Nichols, D.: Robotic approaches for rehabilitation of hand function after stroke. Am. J. Phys. Med. Rehabil. **91**, S242–S254 (2012). https://doi.org/10.1097/PHM.0b013e31826bcedb
20. Babaiasl, M., Mahdioun, S.H., Jaryani, P., Yazdani, M.: A review of technological and clinical aspects of robot-aided rehabilitation of upper-extremity after stroke. Disab. Rehabil. Assist. Technol. **11** (2015). https://doi.org/10.3109/17483107.2014.1002539
21. Kim, G., et al.: Is robot-assisted therapy effective in upper extremity recovery in early stage stroke?—a systematic literature review. J. Phys. Ther. Sci. **29**, 1108–1112 (2017). https://doi.org/10.1589/jpts.29.1108
22. Heo, P., Gu, G.M., Lee, S., Rhee, K., Kim, J.: Current hand exoskeleton technologies for rehabilitation and assistive engineering. Int. J. Precis. Eng. Manuf. **13**, 807–824 (2012). https://doi.org/10.1007/s12541-012-0107-2
23. Piazza, C., et al.: The SoftPro project: synergy-based open-source technologies for prosthetics and rehabilitation. In: Carrozza, M. C., Micera, S., Pons, J.L. (eds.) WeRob 2018. BB, vol. 22, pp. 370–374. Springer, Cham (2019). https://doi.org/10.1007/978-3-030-01887-0_71

Author Index

© The Editor(s) (if applicable) and The Author(s) 2022
C. Biele et al. (Eds.): MIDI 2021, LNNS 440, pp. 299–300, 2022.
https://doi.org/10.1007/978-3-031-11432-8

Printed in the United States
by Baker & Taylor Publisher Services